MONTE CARLO PRINCIPLES
and Neutron Transport Problems

JEROME SPANIER • ELY M. GELBARD

DOVER PUBLICATIONS, INC., MINEOLA, NEW YORK

To Bunny and Helen

Bibliographical Note

This Dover edition, first published in 2008, is an unabridged republication of the work originally published in the "Addison-Wesley Series in Computer Science and Information Processing" by Addison-Wesley Publishing Company, Reading, Massachusetts, in 1969. A new Preface to the Dover edition and an Errata list have been specially prepared for the present edition.

Library of Congress Cataloging-in-Publication Data

Spanier, Jerome, 1930–
   Monte Carlo principles and neutron transport problems / Jerome Spanier, Ely M. Gelbard.
       p. cm.
   "This Dover edition, first published in 2008, is an unabridged republication of the work originally published in the "Addison-Wesley Series in Computer Science and Information Processing" by Addison-Wesley Publishing Company, Reading, Massachusetts, in 1969. A new Preface to the Dover edition and an Errata list have been specially prepared for the present edition."
   Includes index.
   ISBN-13: 978-0-486-46293-6
   ISBN-10: 0-486-46293-5
   1. Monte Carlo method. 2. Neutron transport theory. I. Gelbard, Ely M., 1924– II. Title.

QA298.S663 2008
518'.282—dc22

2007037512

Manufactured in the United States of America
Dover Publications, Inc., 31 East 2nd Street, Mineola, N.Y. 11501

Acknowledgments

This book was written in part at the Bettis Atomic Power Laboratory under the auspices of the Naval Reactors Branch of the United States Atomic Energy Commission. The authors are grateful for the support they received from the Naval Reactors Branch and from Bettis management. In addition, we are indebted to Miss Betty Anderson, Mr. A. E. Dietrich, Mrs. Heid Kuehn, Mrs. Helen Ondis, and Mr. L. A. Ondis II for their patient and careful work in the creation of experimental and production versions of the computer programs MARC and RESQ. Finally, we thank Miss Gail Bell who typed the bulk of the original manuscript and Mrs. Estella Talbert and Mrs. Joanna Altman for their substantial efforts in preparing the manuscript in its final form.

J. S.
E. M. G.

Contents

Preface to the Dover Edition

This book was written at a time when the Monte Carlo method was just beginning to be appreciated as a useful computational method. In the past forty years, this method has become one of the most widely applied techniques available for solving not only problems in neutron transport, but also problems in medicine, finance, operations research, and other fields. My collaboration with Ely Gelbard began in 1955 at Bettis Atomic Power Laboratory and led to the publication by Addison-Wesley of this book in 1969. While our intensive research collaboration lost some strength with my move to the North American Aviation Science Center in 1967 and his to Argonne National Laboratory in 1972, our warm friendship endured until his death in April 2002.

It is a pleasure to record here my gratitude to Ely, who taught me so much about reactor physics in our early years together. Ely's unquenchable curiosity and enthusiasm for tackling the tough problems left a lasting impression on all who had the privilege of knowing him. We took on many such tough problems together in the early days of the nuclear industry in the United States and the methods we developed are still finding application as I write this in 2007. Ely and I were told by many scientists over the years that they learned about Monte Carlo methods by studying our book. The book has been out of print for a number of years and I am grateful to Dover Publications for their willingness to reprint the book, and especially to Senior Editor John Grafton, who spearheaded the effort to publish this Dover edition of *Monte Carlo Principles and Neutron Transport Problems*.

November 2007

JEROME SPANIER
Irvine, California

Introduction

It is possible that some of our readers have had no prior contact with Monte Carlo methods. Some, in fact, may not know what Monte Carlo is and such readers have a right to expect that our Introduction will enlighten them. Certainly it seems appropriate to start a book on Monte Carlo methods with a definition of the term "Monte Carlo." However, we find it difficult to construct a definition which characterizes the Monte Carlo method accurately, completely, and concisely. One can safely say this method, in all its forms, involves some sort of random sampling process. In Monte Carlo computations, samples are drawn from some "parent population" through sampling procedures governed by specified probability laws. Statistical data are collected from the samples, and through analysis of these data one is led to inferences concerning the parent population. But how does one choose the probability laws? And just how does one draw inferences from the data? Here the many variants of Monte Carlo go their separate ways.

In some versions of Monte Carlo, the probability laws are wholly artificial, freely invented, contrived for some special purpose. Thus, for example, wholly artificial Monte Carlo processes have been used to solve the diffusion equation (Ref. 1), and to invert matrices (see Chapter 2 and Refs. 2, 3). Today, however, such Monte Carlo methods are curiosities; though often very interesting, they are rarely very practical.

The use of *completely* artificial probability laws is the most devious of all Monte Carlo strategies. We turn now to the most obvious. Generally, Monte Carlo methods are designed for the study of complicated systems with very many interacting components. Usually, the behavior of the components, as they interact, is governed by known probability laws. In principle it is always possible to incorporate these same laws into the Monte Carlo computational method, so that processes occurring during the course of the computation will be abstract analogs of processes in the real world. If, in Monte Carlo, we sample from probability laws operative in the real world, thus simulating the phenomena we wish to study, the Monte Carlo process is an "analog" or "direct simulation" process.

It will be seen that the ideas underlying analog Monte Carlo are extremely simple and, from a practical point of view, this simplicity is a most attractive feature. However, analog methods also have serious disadvantages. We should not expect that any single method will be efficient and useful under all conceivable circumstances. Indeed we find that analog methods are sometimes very efficient, and sometimes so inefficient as to be virtually worthless. Suppose that we are interested in studying events which, in a given system, occur only very rarely. These events will also rarely occur in an abstract analog of the system. Vast stores of data taken from the analog may, then, contain little information which has value for us. What are we to do when confronted with such a situation? For those who work in Monte Carlo, this is the principal, the most interesting, problem. It is the central problem treated in this book.

Many techniques have been devised to improve the information content in Monte Carlo samples. We shall refer to such techniques as "variance reducing devices." These devices are used to supplement analog methods when analog methods alone are inadequate. No one idea can be said to underlie *all* variance reducing devices but one, perhaps, is especially important. If an event is rare in some given system it will also be rare in any *exact* abstract analog of the system. But it is sometimes possible to work with a distorted analog in which this same event occurs more frequently.† The Monte Carlo sampling laws, in other words, need not be either wholly artificial or completely realistic. Having based them directly on laws of nature we may then choose to distort the sampling laws, to modify them for the sake of efficiency. Of course, this distortion, or "biasing," of the Monte Carlo process complicates the analysis of the sample data. Bias in the data must be entirely removed in the course of the statistical analysis. To mix artificial and analog methods in the same Monte Carlo is a common practice. In fact, a Monte Carlo calculation usually amalgamates many different techniques, both analog and artificial.

Some of the standard variance reducing devices are almost as old as Monte Carlo itself, while some are relatively new. We shall describe many of the old methods in detail for the benefit of readers who know little about Monte Carlo, but we shall also describe two methods which have not yet been widely used. These will be called the "adjoint method" and the "method of superposition." In many respects they are quite similar. We shall see that both lead us to transform given problems into auxiliary problems which are easier to solve. However, the methods differ in that they involve essentially different transformations based, in one case, on the reciprocity relation and, in the other, on the superposition principle. It follows that these methods have somewhat different properties and different ranges of utility.

---

† See Section 3.7, for example.

The adjoint method has been used in thermal neutron studies to compute fluxes at points, and flux averages over very small regions. Of course it is not the only method available for such purposes, but we find it to be a most effective method. In point flux computations it is always necessary to invoke special techniques of some sort since one cannot calculate point values of the flux by analog Monte Carlo. From an academic point of view the problems involved in computing point fluxes and averages over small regions are qualitatively different. It is possible, in principle, to evaluate averages over small regions by analog methods, but the accuracy of analog estimates deteriorates catastrophically as the dimensions of the region shrink to zero. If the region is sufficiently small, the adjoint method will be vastly superior to analog Monte Carlo, while it is not much more difficult to use.

The second method, the superposition method, has been applied to the computation of resonance escape probabilities. We have found that it is very much more efficient than conventional Monte Carlo when the resonance escape probability is close to one. Through the use of superposition it has become possible to treat, by Monte Carlo, isolated resonances and isolated fuel rods. Both the thermal neutron problem and the resonance escape problem will be discussed at length in the second half of this book.

In fact, the book is divided rather sharply into two halves. This division in format developed almost inevitably from the multiplicity of our goals. First of all we hope that the book will be useful as an introduction to Monte Carlo. The first three chapters were written with this goal in mind. There the reader will find an exposition of the general principles of Monte Carlo, developed in the framework of a unified mathematical point of view. The last three chapters are somewhat different in character. In these final chapters we focus directly on the methods of superposition and reciprocity, showing how these methods may be applied to specific problems. Here we confine our attention to neutron transport problems. Our "systems" are configurations of diffusing media containing neutron sources. Our "parent population" is the set of all possible neutron histories, and our "samples" are histories drawn from this set. Chapters 5 and 6 are entirely given over to the study of two Monte Carlo problems, i.e. the computation of thermal neutron fluxes and resonance escape probabilities. In these final chapters we develop the tricks and techniques which we need to make the adjoint and superposition methods practical, and we analyze the capabilities of both.

That these methods are practical we have no doubt. We have worked with them; we have seen them used extensively by reactor physicists and nuclear designers. In fact, throughout this book we confine our attention to Monte Carlo methods with which we have had close contact. Whatever the merits of such an approach, it has tended to limit the scope of the book. We shall not discuss the special techniques which have been applied to shielding calculations, nor shall we say anything about criticality calculations.

Therefore, we cannot claim to have written a general treatise on Monte Carlo. Nevertheless, we hope that our first three chapters are broad enough in scope to constitute a satisfactory *introduction* to Monte Carlo.

The first three chapters are self-contained in the sense that we have based our exposition on elementary mathematical concepts. We require of the reader only those mathematical skills which are standard equipment for most engineers and scientists and, in addition, some slight familiarity with the transport equation. Again we must point out that the last three chapters are rather different from the first. These last chapters are directed to the reader who has a special interest in reactors and has had some contact with the problems and techniques of reactor theory. It will be assumed, for example, that he knows a little about the properties of thermal neutrons, that he knows about resonances, that he knows what is meant by a "fuel rod," etc. On the other hand, all the transport theory which is essential to our work will be developed as we need it.

We conclude these preliminary remarks with comments on the use of Monte Carlo as a reactor design tool. Ten years ago we would not have referred to Monte Carlo as a design tool. It was, so very recently, an awkward, expensive computational standard, and little more. But, as computers grow in power, it becomes necessary over and over again to revise one's point of view. Consider, for example, the capabilities of the Philco-2000-212. At the Bettis Atomic Power Laboratory, a typical Monte Carlo thermal utilization calculation runs two or three minutes on this machine. A typical superposition resonance escape calculation takes about the same amount of time. Yet today the Philco-2000-212 already belongs to a past generation of computers. It is very difficult to tell where the evolution of computers will lead us in the next ten years, but we think there is good reason to expect that Monte Carlo will become increasingly important. If computers continue to grow, more accurate, more realistic, and therefore more complicated, reactor computations will become feasible. And, as problems become more complicated, the advantages of Monte Carlo over analytic methods become more and more impressive.

REFERENCES

1. John H. Curtiss, "Sampling Methods Applied to Differential and Difference Equations," *Seminar on Scientific Computation, Proceedings, New York*, IBM Corp. (November, 1949).
2. G. E. Forsythe and R. A. Leibler, "Matrix Inversion by a Monte-Carlo Method," *Math. Tables Aids to Comp.*, 4 (31), 127 (1950).
3. W. Wasow, "A Note on the Inversion of Matrices by Random Walks," *Math. Tables Aids to Comp.*, 6, 78 (1952).

1

Fundamentals of Monte Carlo Methods

1.1 INTRODUCTION

The single most important, and most distinctive, feature of Monte Carlo is the central role of random sampling methods. In all Monte Carlo calculations it is necessary to draw samples from specified probability distributions. In this first chapter we shall study the elements of Monte Carlo sampling theory. Some basic definitions are introduced in Section 1.2, while in Sections 1.3 and 1.4 we develop as much probability theory as we require in our work. The next four sections deal specifically with several of the more widely used sampling techniques, techniques used to construct samples with given distribution functions. Finally, in Section 1.9, we discuss the most fundamental of all sampling problems, i.e. the problem of generating random numbers.

1.2 RANDOM VARIABLES

Monte Carlo methods are probabilistic in nature; a proper understanding of them, therefore, requires some acquaintance with probability theory. For this reason we begin with a brief discussion of some elementary topics in probability theory. Two excellent references for further reading in probability theory are Halmos (Ref. 1) and Loève (Ref. 2).

The first and most basic notion encountered is that of an *event*, by which we mean the occurrence of a specified outcome of an experiment. For example, in an experiment consisting of the throw of a single die the occurrence of any of the 6 faces, say face A, is an event. It is mathematically convenient to be able to combine events to obtain other events; e.g. the simultaneous occurrence of face A and face B ($A \neq B$) is an event, the impossible event, denoted ø. Events will be denoted by capital Greek letters $\Lambda, \Omega, \ldots$. As we shall see, elementary laws of composition of sets are useful in discussing classes of events.

From the outset we shall postulate the existence of the certain event Ω, the event which includes all possible events and, therefore, which always occurs. In symbols this is expressed by $\Lambda \subset \Omega$ for every event Λ, and is read

"Λ is contained in Ω." In the example of the die, Ω would be represented as a set with 6 elements, one for each face. To every event Λ there corresponds a complementary event, Λ', which occurs if and only if Λ does not occur. The event Λ', called the complement of Λ, consists of all elements of Ω not in Λ. Thus, $\Omega' = \emptyset$ the null event or impossible event.

Events may be combined into other events by means of elementary set-theoretic operations. The event $\Lambda \cap \Gamma$, called the intersection of Λ and Γ, consists of all elements common to both Λ and Γ. The event $\Lambda \cap \Gamma$ occurs if and only if both Λ and Γ occur. If $\Lambda \cap \Gamma = \emptyset$ we say that Λ and Γ are disjoint; intuitively, Λ and Γ are mutually exclusive, or incompatible, events. For example, $\Lambda \cap \Lambda' = \emptyset$ for all events Λ. The event $\Lambda \cup \Gamma$, called the union of Λ and Γ, consists of all elements in either Λ or Γ or both. The event $\Lambda \cup \Gamma$ occurs if and only if either Λ or Γ or both occur. More than two events may be combined by means of these relations: If $\Lambda_1, \Lambda_2, \ldots, \Lambda_n$ are events, we use the symbol $\bigcap_{i=1}^{n} \Lambda_i$ instead of $\Lambda_1 \cap \Lambda_2 \cap \cdots \cap \Lambda_n$ to denote the simultaneous occurrence of all the events Λ_i, and $\bigcup_{i=1}^{n} \Lambda_i$ instead of $\Lambda_1 \cup \Lambda_2 \cup \cdots \cup \Lambda_n$ to denote the occurrence of any of the events Λ_i.

The elementary laws of set theory apply to these operations and yield familiar relationships, for example

$$(\Lambda \cap \Gamma)' = \Lambda' \cup \Gamma', \qquad (\Lambda \cup \Gamma)' = \Lambda' \cap \Gamma'. \qquad (1.2.1)$$

We shall not list all of the many possible relationships among events. Again, the interested reader may consult Halmos (Ref. 1) or Loève (Ref. 2) for details. In the investigation of outcomes of an experiment, all countable combinations of events which may be formed by the operations ', $\cap$, $\cup$ are also postulated to be events associated with the given experiment.

At this point we digress briefly in order to clarify what we shall mean by an experiment. Since our object in this book is to discuss numerical experiments, all of whose conditions and probability laws shall be taken as axiomatic and under our control, our formulation of the definition of experiment will be directed toward this application. In this way we are able to avoid the philosophical difficulties which ordinarily accompany an attempt to describe the set of fixed conditions and variable conditions (e.g. temperature, pressure, physical dimensions) which characterize a real experiment. Thus we define an abstract probability model as consisting of various fixed conditions and probability laws. Such a model, which is purely a mathematical construct, may be based directly on a physical system or it may be wholly or partly contrived to suit a specific purpose. In Chapter 2 we discuss analog random walk processes, which are examples of abstract probability models based directly on the behavior of neutrons governed by a steady-state transport equation. Non-analog random walk processes, on the other hand, are abstract probability models created in such a way that certain integral properties of the steady-state flux distribution are preserved. The detailed behavior of the neutrons, however, is not the same as in the analog case. An even more contrived model is the one used in Chapter 2 to solve matrix equations. Artificial models have also been used to obtain Monte Carlo solutions of the diffusion equation (see Ref. 1 of the Introduction).

By an idealized experiment we shall mean a simulation, by numerical means, of an abstract probability model. The precise way in which this simulation takes place will be clarified in Sections 1.5–1.9, and in Chapter 2. For the moment we wish only to delineate the class of problems (idealized experiments) at which the discussion in this book is aimed.

To illustrate, if the real experiment were the throw of a die, in our abstract probability model we would choose to ignore certain physical conditions such as weight of the die, dimensions of the die faces, etc. We would characterize the idealized experiment completely by one probability law which associates the probability 1/6 with each of the 6 faces. In a more complicated idealized experiment, such as the random walk of a particle, the fixed conditions would consist of the geometry and cross sections in the energy range under consideration. The probability laws govern, for example, the choice of distances travelled between collisions, the choice between scattering and absorption upon collision, and the other random elements associated with the process. Because our experiment is numerical rather than physical we may take both the fixed conditions and probability laws as axiomatic and make no attempt to justify them on physical grounds.

The intuitive notion of probability has to do with the frequency of occurrence of an event Λ in repeated independent trials of an experiment. In order to give this intuitive notion precision, the *probability* is defined as a real-valued function on the events of an experiment satisfying

1) $P(\emptyset) = 0, \quad P(\Omega) = 1, \qquad 0 \leq P(\Lambda) \leq 1$ for every event Λ;

2) $P\left(\bigcup_{i=1}^{\infty} \Lambda_i\right) = \sum_{i=1}^{\infty} P(\Lambda_i), \qquad \text{if} \quad \Lambda_i \cap \Lambda_j = \emptyset, \quad i \neq j.$ (1.2.2)

We point out that this abstract notion of probability of an event is more general than one might expect based on intuitive grounds. If an experiment is repeated n times and if the event Λ occurs $n(\Lambda)$ times, one expects the ratios $n(\Lambda)/n$ to cluster about a *unique* number, the probability of Λ. We require only that the function P assign to every event Λ a number with the properties (1.2.2). This flexibility is needed to deal with the possibility that probabilities of events may not come directly from simulation of a physical process.

The probability is a real-valued function on certain subsets of Ω, which we call the events of Ω. Certain real-valued functions on the points of Ω, called random variables, will also be important in our analysis of an idealized experiment. To define this concept, let a point of Ω be denoted by the small

Greek letter ω and let ξ be a real-valued function on the points of Ω. Let $\Lambda(t) = \{\omega : \xi(\omega) \leq t\}$, by which we mean "the set of all points ω such that $\xi(\omega) \leq t$." Notice that $\Lambda(t)$ is a subset of Ω which depends on the real number t. If for every t the set $\Lambda(t)$ is an event, then the function ξ is called a *random variable* on Ω. The quantities of real interest in an experiment are usually studied by means of such random variables.

If ξ is a random variable on Ω, then the probability

$$P(\Lambda(t)) = P\{\omega \mid \xi(\omega) \leq t\}$$

is defined for every t. We shall frequently write $P\{\xi \leq t\}$ in place of $P(\Lambda(t))$. The real-valued function of a real variable defined by

$$F(t) = P\{\xi \leq t\} \tag{1.2.3}$$

is called the *distribution function*, or sometimes simply *distribution*, of the random variable ξ. The distribution F has several important properties which are easily deduced.

1) F is continuous on the right at every t.
2) F is monotone nondecreasing.
3) $F(-\infty) = 0$, $F(\infty) = 1$.
4) $F(b) - F(a) = P\{a < \xi \leq b\}$, $a < b$.
5) If t_0 is a point of discontinuity of F with a jump of height p, then $P\{\xi = t_0\} = p$ and there is a nonzero probability that the random variable takes on the value t_0. If the derivative $F'(t) = f(t)$ exists at a point t, then

$$\lim_{\Delta \to 0} P(t - \Delta/2 < \xi \leq t + \Delta/2) = f(t)\Delta.$$

The function $f(t) = F'(t)$ (if it exists for all t) will be called the *density function* of the random variable ξ or of the distribution F.

To illustrate the preceding discussion we again consider the experiment which consists in throwing a single die. The certain event Ω may be regarded as a set consisting of 6 points $\omega_1, \omega_2, \ldots, \omega_6$ corresponding to the 6 faces of the die. All possible subsets of Ω form the events of the experiment and the natural probability function P is defined by $P\{\omega_i\} = 1/6$, $1 \leq i \leq 6$. The set $\{\omega_1\} \cup \{\omega_2\}$ corresponds to the event "either a one or a two" on a single throw, while $\{\omega_1\} \cap \{\omega_2\} = \emptyset$ corresponds to the fact that both a one and a two cannot appear in a single throw. Define a random variable ξ on Ω by $\xi(\omega_i) = i$, $1 \leq i \leq 6$. Since every subset of Ω is an event of Ω, every real-valued function on Ω defines a random variable. For example, $\{\omega \mid \xi(\omega) \leq 3\} = \{\omega_1, \omega_2, \omega_3\}$ is a well-defined event whose probability may be calculated using (1.2.2) by

$$P\{\omega_1, \omega_2, \omega_3\} = P\{\omega_1\} + P\{\omega_2\} + P\{\omega_3\} = 1/2,$$

since the ω_i are disjoint.

A random variable ξ and its distribution F will be called *discrete* if the distribution function $F(x)$ is a step function with a finite number of jump discontinuities in any finite interval. Thus, it is completely described by a sequence $x_1, x_2, \ldots$ of points and a sequence $p_1, p_2, \ldots$ of probabilities such that $P\{\xi = x_i\} = p_i$, $i = 1, 2, \ldots$, and $\sum_i p_i = 1$. The distribution $F(x)$ has the form

$$F(x) = \sum_{x_i \leq x} p_i. \tag{1.2.4}$$

In contrast to the discrete case, a random variable ξ and its distribution F will be called *continuous* if the density function $F'(x) = f(x)$ exists everywhere and is continuous except possibly at a finite number of points in any finite interval. If F is continuous, we have

$$F(x) = \int_{-\infty}^{x} f(t)\, dt. \tag{1.2.5}$$

Most random variables occurring in practice are either of the discrete or the continuous type and we shall restrict our attention to these two types. Clearly, any continuous distribution can be arbitrarily closely approximated by a discrete distribution.

An important example of a continuous distribution is provided by the random variable ξ whose density function $f(x)$ is defined by

$$f(x) = \begin{cases} \dfrac{1}{b - a} & \text{for } a < x < b, \\ 0 & \text{otherwise.} \end{cases} \tag{1.2.6}$$

Such a random variable is said to be uniformly distributed in the interval (a, b). Its distribution function, $F(x)$, is

$$F(x) = \begin{cases} 0 & \text{for } x \leq a, \\ \dfrac{x - a}{b - a} & \text{for } a < x < b, \\ 1 & \text{for } x \geq b. \end{cases} \tag{1.2.7}$$

Another important density function is that of a normal variable

$$\phi(x) = \frac{1}{\sqrt{2\pi}} e^{-x^2/2}, \qquad -\infty \leq x \leq \infty \tag{1.2.8}$$

which has the distribution function

$$\Phi(x) = \frac{1}{\sqrt{2\pi}} \int_{-\infty}^{x} e^{-t^2/2}\, dt, \qquad -\infty \leq x \leq \infty. \tag{1.2.9}$$

This distribution is of fundamental importance in most of the applications of probability theory and will be encountered again in later sections.

Frequently, two or more random variables are simultaneously observed in the performance of a single experiment. This fact necessitates a generalization of the preceding concepts to many variables; for simplicity we treat only two. Let ξ, η be random variables associated with a certain experiment. Then the probability that simultaneously ξ takes on values $\leq s$ and η takes on values $\leq t$ defines the *joint distribution function* of ξ and η:

$$F(s, t) = P\{\xi \leq s, \eta \leq t\}. \tag{1.2.10}$$

This function has properties analogous to one-dimensional distributions—we shall not bother to enumerate them here. The distribution $F(s, t)$ will be called *discrete* if it is discrete in each variable; i.e. if F is a step function with a finite number of jump discontinuities in any finite interval for either variable. If we let

$$p_{ij} = P\{\xi = x_i, \eta = y_j\}, \quad i, j = 1, 2, \ldots \tag{1.2.11}$$

be the probability that simultaneously ξ takes on the value x_i and η takes on the value y_j, then

$$F(x, y) = \sum_{\substack{x_i \leq x \\ y_j \leq y}} p_{ij} \tag{1.2.12}$$

and

$$\sum_{i,j} p_{ij} = 1.$$

If we further denote by p_i the probability that $\xi = x_i$ without regard for the value of η, then

$$p_i = \sum_j p_{ij} \tag{1.2.13}$$

and the function

$$F_1(x) = \sum_{x_i \leq x} p_i \tag{1.2.14}$$

defines a discrete distribution function, the *marginal distribution* of ξ associated with $F(x, y)$. The marginal distribution of η is defined analogously.

The two-dimensional distribution $F(x, y)$ is called *continuous* if its density function

$$f(x, y) = \frac{\partial^2 F(x, y)}{\partial x \, \partial y}$$

exists everywhere and is continuous except possibly in certain curves of the plane. In this case we have

$$F(x, y) = \int_{-\infty}^{y} \int_{-\infty}^{x} f(s, t) \, ds \, dt. \tag{1.2.15}$$

Similarly, the *marginal distribution* of ξ is defined by

$$F_1(x) = P(\xi \leq x) = \int_{-\infty}^{x} \left[\int_{-\infty}^{\infty} f(s, t)\, dt \right] ds \qquad (1.2.16)$$

and the *marginal density of* ξ is

$$f_1(x) = F_1'(x) = \int_{-\infty}^{\infty} f(x, t)\, dt. \qquad (1.2.17)$$

Consider now two random variables ξ, η having a joint distribution $F(x, y)$. If

$$P\{a_1 < \xi \leq b_1, a_2 < \eta \leq b_2\} = P\{a_1 < \xi \leq b_1\}P\{a_2 < \eta \leq b_2\}$$
$$(1.2.18)$$

for all a_1, b_1, a_2, b_2, we say that ξ and η are *independent* random variables. Letting a_1, a_2 approach $-\infty$ we find that for independent random variables

$$F(x, y) = F_1(x)F_2(y), \qquad (1.2.19)$$

where F_1 and F_2 are the two marginal distributions. If the density functions exist, it can also be shown that

$$f(x, y) = f_1(x)f_2(y), \qquad (1.2.20)$$

where f_1 and f_2 are the marginal density functions. Any of these conditions can be shown to be necessary and sufficient for independence. For functions of n variables the conditions are

$$F(x_1, \ldots, x_n) = F_1(x_1) \cdots F_n(x_n) \qquad (1.2.21)$$

or

$$f(x_1, \ldots, x_n) = f_1(x_1) \cdots f_n(x_n) \qquad (1.2.22)$$

if the densities exist. Thus, it is necessary that each random variable be independent of all the others.

As an example of a continuous two-dimensional density function consider

$$f(x, y) = \begin{cases} \dfrac{1}{(b_1 - a_1)(b_2 - a_2)} & \text{for } a_1 < x < b_1, a_2 < y < b_2, \\ 0 & \text{otherwise.} \end{cases} \qquad (1.2.23)$$

This is the two-dimensional generalization of the uniform distribution defined by Eq. (1.2.6). If we let ξ and η be the random variables involved, then it is easily seen that the marginal distributions of ξ and η are each uniform and, further, that ξ and η are independent random variables.

Let ξ, η be two random variables, assumed continuous for the purpose of the present discussion, with joint density function $f(x, y)$. We have seen that the marginal density functions of ξ, η are

$$f_1(x) = \int_{-\infty}^{\infty} f(x, y) \, dy \tag{1.2.24}$$

$$f_2(y) = \int_{-\infty}^{\infty} f(x, y) \, dx. \tag{1.2.25}$$

Define a new function by

$$f(y \mid x) = \frac{f(x, y)}{f_1(x)}. \tag{1.2.26}$$

This function, which has all the properties of a density function for each fixed x such that $f_1(x) \neq 0$, is called the *conditional density* of η relative to the hypothesis that $\xi = x$. Thus, we have

$$\int_{-\infty}^{\infty} f(y \mid x) \, dy = \frac{\int_{-\infty}^{\infty} f(x, y) \, dy}{\int_{-\infty}^{\infty} f(x, y) \, dy} = 1 \tag{1.2.27}$$

and $f(y \mid x) \geq 0$. By interchanging the variables ξ and η we obtain the conditional density of ξ relative to the hypothesis that $\eta = y$:

$$f(x \mid y) = \frac{f(x, y)}{f_2(y)}. \tag{1.2.28}$$

We see that, according to the definition of independence (Eq. 1.2.20), if ξ and η are independent random variables, then

$$f(x \mid y) = f_1(x) \qquad \text{and} \qquad f(y \mid x) = f_2(y). \tag{1.2.29}$$

Conversely, if $f(x \mid y) = p(x)$ is independent of y, then

$$f(x, y) = p(x) f_2(y) \tag{1.2.30}$$

and we conclude that ξ and η are independent.

In similar fashion we define the conditional density of y relative to the hypothesis that $x < x_0$ by

$$f(y \mid x < x_0) = \int_{-\infty}^{x_0} f(y \mid x) \, dx = \frac{\int_{-\infty}^{x_0} f(x, y) \, dx}{\int_{-\infty}^{x_0} f_1(x) \, dx}. \tag{1.2.31}$$

Associated with each random variable ξ is its *mean* or *expected value*, denoted $E[\xi]$. In the general case, if F is the distribution function of ξ, the expected value of ξ may be defined by

$$E[\xi] = \int_{-\infty}^{\infty} x \, dF(x), \tag{1.2.32}$$

where the integral is taken in the Stieltjes sense. If ξ is discrete, this integral becomes

$$E[\xi] = \sum_{i=1}^{\infty} p_i x_i, \tag{1.2.33}$$

while if ξ is continuous,

$$E[\xi] = \int_{-\infty}^{\infty} x f(x) \, dx. \tag{1.2.34}$$

We shall say that the mean value exists provided that the sum or integral which represents it is absolutely convergent.† In the example of the die cited earlier,

$$E[\xi] = \sum_{i=1}^{6} i \cdot \tfrac{1}{6} = 3.5, \tag{1.2.35}$$

which illustrates the fact that the expected value need not even be taken on by the random variable ξ. As an example of a random variable ξ whose expected value does not exist, consider the Cauchy distribution whose density function is defined by

$$f(x) = \frac{1}{\pi(1 + x^2)}, \qquad -\infty \leq x \leq \infty. \tag{1.2.36}$$

Since the integral

$$E[\xi] = \frac{1}{\pi} \int_{-\infty}^{\infty} \frac{x}{1 + x^2} \, dx \tag{1.2.37}$$

is not absolutely convergent,‡ no mean value exists for this distribution.

If ξ is any random variable and ϕ any function, then $\eta = \phi(\xi)$ is also a random variable. Its mean value $E[\eta]$, provided it exists, may be calculated by the formula

$$E[\eta] = \int_{-\infty}^{\infty} \phi(x) \, dF(x), \tag{1.2.38}$$

where $F(x)$ is the distribution of ξ.

---

† One often builds up random variables ξ by taking linear combinations of elementary random variables ξ_i, defined simply by

$$\xi_i(\omega) = \begin{cases} 1, & \omega \in \Lambda_i, \\ 0, & \omega \notin \Lambda_i. \end{cases}$$

Absolute convergence is needed to guarantee that the expectation of such a sum shall be independent of the order of the terms and, hence, independent of the order of occurrence of the Λ_i.

‡ The limit

$$\lim_{M \to \infty} \int_{-M}^{M} f(x) \, dx,$$

if it exists, is often called the Cauchy principal value of the improper integral $\int_{-\infty}^{\infty} f(x) \, dx$. The integral (1.2.37), then, does have a Cauchy principal value of 0. There seems to be no unanimity of terminology used in mathematics texts in describing convergence of improper integrals. The absolute convergence we require is equivalent to the much stronger condition that the limit $\lim \int_{-M}^{N} f(x) \, dx$ exist as M and N independently tend to infinity.

From the definition (1.2.32) it may easily be shown that

$$E[a\xi + b] = aE[\xi] + b, \tag{1.2.39}$$

where a and b are constant, and that

$$E[\xi + \eta] = E[\xi] + E[\eta] \tag{1.2.40}$$

for random variables ξ and η. A second important result is that for ξ, η independent random variables,

$$E[\xi \cdot \eta] = E[\xi]E[\eta]. \tag{1.2.41}$$

This is a direct consequence of the condition (1.2.20) for independence.

If ξ is a random variable and n a nonnegative integer, the expected value

$$\alpha_n \equiv E[\xi^n] = \int_{-\infty}^{\infty} x^n \, dF(x) \tag{1.2.42}$$

is called the *n-th moment* of ξ. If we denote the mean by $m = E[\xi]$, then the moments

$$\mu_n = E[(\xi - m)^n] \tag{1.2.43}$$

are called *central moments* of ξ and are particularly important in applications. Especially important is the second central moment, called the *variance*, μ_2, which is used to measure the dispersion of the distribution about its mean value m. The variance μ_2 is often denoted by $\sigma^2[\xi]$ and its positive square root, denoted $\sigma[\xi] = \sqrt{\mu_2}$, is called the *standard deviation* of ξ.

As an illustration of the significance of the standard deviation as a measure of the dispersion of a distribution we cite the following result due to Tchebycheff (Ref. 3).

Theorem 1.1 Let ξ be a random variable whose expected value m and standard deviation σ exist and let $k > 0$. Then

$$P\{|\xi - m| > k\sigma\} \leq 1/k^2. \tag{1.2.44}$$

Since Theorem 1.1 is free of assumptions about the distribution of ξ other than the existence of m and σ, it does not, in general, provide sharp bounds on the rate of decrease of mass of the distribution at large distances from the mean. To obtain sharper bounds, more information about the distribution is needed. We shall discuss this further in Section 1.3.

The definition of expectation for multi-dimensional distributions is straightforward: if ξ, η are random variables with joint distribution $F(x, y)$ and $\phi(x, y)$ is any function of two variables, then

$$E[\phi(\xi, \eta)] = \int_{-\infty}^{\infty} \int_{-\infty}^{\infty} \phi(x, y) \, dF(x, y), \tag{1.2.45}$$

provided the sum or integral is absolutely convergent.

In similar fashion, conditional mean values are defined in an obvious way. Suppose, for example, that ξ and η are continuous random variables with joint density function $f(x, y)$. Then, according to Eq. (1.2.26), the conditional expected value of η relative to the hypothesis that $\xi = x$ is

$$E[\eta \mid \xi = x] = \int_{-\infty}^{\infty} yf(y \mid x) \, dy$$

$$= \frac{\int_{-\infty}^{\infty} yf(x, y) \, dy}{\int_{-\infty}^{\infty} f(x, y) \, dy}. \tag{1.2.46}$$

1.3 STATISTICAL ANALYSIS

In this section we present an outline of the theory on which, later, we shall base the rigorous statistical analysis for a general Monte Carlo program.

Consider an experiment with only two possible outcomes, an event which may occur with probability p and its complement. In the terminology introduced in Section 1.2, this would be described by saying that the certain event Ω is a space consisting of two points, ω and ω', with a probability P on Ω defined by

$$P(\omega) = p, \qquad 0 \leq p \leq 1,$$
$$P(\omega') = q = 1 - p. \tag{1.3.1}$$

Further define a random variable ξ on the space Ω by

$$\xi(\omega) = 1, \qquad \xi(\omega') = 0. \tag{1.3.2}$$

The probability distribution of ξ is defined by

$$F(t) = P\{\xi \leq t\} = \begin{cases} 0 & \text{if } t < 0, \\ q & \text{if } 0 \leq t < 1, \\ 1 & \text{if } t \geq 1. \end{cases} \tag{1.3.3}$$

The function $F(t)$ can be written

$$F(t) = p\chi(t - 1) + q\chi(t), \tag{1.3.4}$$

where

$$\chi(t) = \begin{cases} 0 & \text{if } t < 0. \\ 1 & \text{if } t \geq 0. \end{cases} \tag{1.3.5}$$

Intuitively, the experiment has two possible outcomes, ω and ω', occurring with probabilities p and q, respectively, and the random variable ξ counts one if ω occurs, zero if not. The discrete random variable ξ is the simplest example one can consider but is extremely important in applications.

If N independent trials of the experiment are performed, the appropriate probability space to examine is the space Ω^N consisting of all N-tuples $(\omega_1, \ldots, \omega_N)$ of points of Ω. Further, the appropriate probability on Ω^N is the product probability defined by

$$P_N(\omega_1, \ldots, \omega_N) = \prod_{i=1}^{N} P(\omega_i). \tag{1.3.6}$$

Define N random variables ξ_i on Ω^N by

$$\xi_i(\omega_1, \ldots, \omega_N) = \xi(\omega_i), \qquad 1 \leq i \leq N, \tag{1.3.7}$$

so that each ξ_i is a discrete random function of N variables. Then the sum

$$\xi^{(N)} = \sum_{i=1}^{N} \xi_i \tag{1.3.8}$$

is also a random variable on Ω^N representing the total number of occurrences of the event ω in N repetitions of the basic experiment. It is also a discrete variable, taking on the possible values $0, 1, 2, \ldots, N$, and its probability distribution is defined by

$$F_N(t) = P\{\xi^{(N)} \leq t\}$$

$$= \sum_{k \leq t} \binom{N}{k} p^k q^{N-k}. \tag{1.3.9}$$

Here $\binom{N}{k}$ is the binomial coefficient and the sum is over all integers $k \leq t$: the distribution F_N is called the binomial distribution. Using the definitions and properties developed in Section 1.2, one can verify that

$$E[\xi^{(N)}] = Np, \qquad \sigma[\xi^{(N)}] = \sqrt{Npq}. \tag{1.3.10}$$

We have already introduced the normal distribution (Eqs. 1.2.8 and 1.2.9) defined by

$$\Phi(x) = \frac{1}{\sqrt{2\pi}} \int_{-\infty}^{x} e^{-t^2/2} \, dt, \qquad -\infty \leq x \leq \infty. \tag{1.3.11}$$

This distribution is of continuous type, its density function being

$$\phi(x) = \Phi'(x) = \frac{1}{\sqrt{2\pi}} e^{-x^2/2}. \tag{1.3.12}$$

It may be shown that the random variable ξ associated with Φ has zero expectation and unit standard deviation,

$$E[\xi] = 0, \qquad \sigma[\xi] = 1. \tag{1.3.13}$$

A random variable ξ will be called *normal* (m, σ) (with mean m, standard deviation σ) if it has as its distribution function the function

$$\Phi\left(\frac{t - m}{\sigma}\right), \qquad \sigma > 0.$$

From the properties developed in Section 1.2 it is easy to show that if ξ is normal (m, σ), then $(\xi - m)/\sigma$ is normal $(0, 1)$.

Returning now to the binomial distribution, we interpreted $\xi^{(N)}$ as the total number of occurrences of the event ω in N independent trials of the binomial experiment. Then

$$\bar{\xi}^{(N)} = \frac{\xi^{(N)}}{N} \tag{1.3.14}$$

measures the frequency of occurrence of the event ω and is sometimes referred to as the success ratio. Now, using (1.2.39), (1.2.40), and (1.2.41), we find that

$$E[\bar{\xi}^{(N)}] = p, \qquad \sigma[\bar{\xi}^{(N)}] = \sqrt{\frac{pq}{N}}. \tag{1.3.15}$$

Let $\epsilon > 0$ be arbitrary and let $k = \epsilon\sqrt{N/pq}$. Then, if we apply Tchebycheff's theorem (Theorem 1.1 of Section 1.2) to the random variable $\bar{\xi}^{(N)}$, we obtain

$$P\{|\bar{\xi}^{(N)} - p| > \epsilon\} < \frac{pq}{N\epsilon^2} \leq \frac{1}{4N\epsilon^2}. \tag{1.3.16}$$

By choosing N sufficiently large, the right-hand side becomes arbitrarily small and we have

Theorem 1.2 (Bernoulli) For any given $\epsilon > 0$, the probability that the success ratio $\bar{\xi}^{(N)}$ differs in magnitude from p by an amount exceeding ϵ tends to zero as N tends to infinity.

Bernoulli's theorem, although useful for making rigorous certain intuitive ideas about probability, unfortunately tells us nothing about the size of the error involved in accepting the success ratio as an estimate of the probability p. Of course, Tchebycheff's theorem, which was used in the proof of Bernoulli's theorem, does provide a weak bound for the error. In order to sharpen this error bound, we must go to a somewhat deeper result, the central limit theorem, which we now discuss.

Let $\xi_1, \ldots, \xi_N$ be N independent and normally distributed random variables. Then it may be shown that the variable

$$\xi = \sum_{i=1}^{N} \xi_i \tag{1.3.17}$$

is also normally distributed. The central limit theorem states essentially that, under very general conditions, ξ will be approximately normally distributed even if the ξ_i are not normal. We shall state one form of this theorem as follows (see, for example, Ref. 3).

Theorem 1.3 Let $\xi_1, \xi_2, \ldots$ be a sequence of independent and identically distributed random variables with common mean m and standard deviation σ. Then the average

$$\bar{\xi}_N = \frac{1}{N} \sum_{i=1}^{N} \xi_i \qquad (1.3.18)$$

is asymptotically normal $(m, \sigma/\sqrt{N})$; i.e.

$$\lim_{N \to \infty} P \left\{ \frac{\bar{\xi}_N - m}{\sigma/\sqrt{N}} \leq x \right\} = \Phi(x)$$

$$= \frac{1}{\sqrt{2\pi}} \int_{-\infty}^{x} e^{-t^2/2} \, dt. \qquad (1.3.19)$$

This remarkable theorem will presently be applied to the results of a typical Monte Carlo calculation. First, however, we point out that the theorem, as we have stated it, assumes that m and σ exist; i.e. are given by absolutely convergent integrals. Without such assumptions the conclusion of the theorem might be invalid. For example, let ξ_1 and ξ_2 be two independent random variables, each having a Cauchy distribution (see Section 1.2) with density function

$$f(x) = \frac{1}{\pi(1 + x^2)}, \qquad -\infty \leq x \leq \infty. \qquad (1.3.20)$$

Then the density function of $\xi_1 + \xi_2$ is

$$g(x) = \frac{1}{\pi^2} \int_{-\infty}^{\infty} \frac{dt}{[1 + (x - t)^2][1 + t^2]}$$

$$= \frac{2}{\pi(4 + x^2)}. \qquad (1.3.21)$$

The average, $(\xi_1 + \xi_2)/2$, has the density function

$$2g(2x) = \frac{1}{\pi(1 + x^2)}, \qquad (1.3.22)$$

which is that of the Cauchy distribution again. By induction, if $\xi_1, \xi_2, \ldots, \xi_N$ are N independent random variables, each having a Cauchy distribution, then the arithmetic mean $\bar{\xi}_N$ also has the Cauchy distribution for any N, in violation of the conclusion of the central limit theorem. This happens, of course, because the mean of the Cauchy distribution does not exist.

Now let us see how the central limit theorem might be applied to a Monte Carlo calculation. Suppose particles undergo random walks according to probability laws which we shall specify in later sections. Let the set of all possible random walks be denoted Ω. Then we would like to associate a point ω of Ω with a weight t, a real number, which represents the contribution of the random walk ω to the particular quantity being estimated. This idea is made precise by defining a random variable ξ on Ω and by regarding the values $t = \xi(\omega)$ as the weights in question. Let the random variable ξ have the distribution

$$F(t) = P\{\xi \leq t\}, \tag{1.3.23}$$

with mean

$$m = E[\xi] = \int_{-\infty}^{\infty} t \, dF(t) \tag{1.3.24}$$

and variance

$$\sigma^2 = E[(\xi - m)^2] = \int_{-\infty}^{\infty} (t - m)^2 \, dF(t) \tag{1.3.25}$$

assumed finite. If K independent random walks ω_i, $1 \leq i \leq K$ are followed, K random variables ξ_i may be defined on the space Ω^K of all K-tuples of points of Ω by

$$\xi_i(\omega_1, \ldots, \omega_K) = \xi(\omega_i), \qquad 1 \leq i \leq K. \tag{1.3.26}$$

The random variables ξ_i are independent and identically distributed; thus, the hypotheses of Theorem 1.3 are satisfied and allow us to conclude that

$$\bar{\xi}_K = \frac{1}{K} \sum_{i=1}^{K} \xi_i \tag{1.3.27}$$

is asymptotically normal $(m, \sigma/\sqrt{K})$. The random variable $\bar{\xi}_K$ represents the average weight of K particles performing independent random walks and the theorem states that this average weight approximates m if K is sufficiently large. Let $\sigma_K = \sigma/\sqrt{K}$. Then, since $\bar{\xi}_K$ is asymptotically (for large K) normal (m, σ_K), it follows that $(\bar{\xi}_K - m)/\sigma_K$ is asymptotically normal $(0, 1)$. Thus, for K sufficiently large, the following asymptotic equalities hold:

$$P\left\{ \left| \frac{\bar{\xi}_K - m}{\sigma_K} \right| < \epsilon' \right\} \sim \frac{1}{\sqrt{2\pi}} \int_{-\infty}^{\epsilon'} e^{-t^2/2} \, dt - \frac{1}{\sqrt{2\pi}} \int_{\epsilon'}^{\infty} e^{-t^2/2} \, dt, \tag{1.3.28}$$

where the symbol $\sim$ is read "is asymptotically equal to." After rearrangement we are led to the following statement: given any $\epsilon > 0$,

$$P\{|\bar{\xi}_K - m| < \epsilon\} \sim \sqrt{\frac{2}{\pi}} \int_0^{\epsilon/\sigma_K} e^{-t^2/2} \, dt. \tag{1.3.29}$$

Since the right-hand side approaches one as K approaches infinity (because $\sigma_K \to 0$), we are assured that $\bar{\xi}_K$ tends to m in the sense of Eq. (1.3.29) and the probability that ξ_K will deviate from m by more than a given amount ϵ can be estimated.

1.4 SAMPLING DISTRIBUTIONS AND CONFIDENCE INTERVALS

Let Ω be the space of points associated with some experiment and let ξ be a random variable on Ω. As we have already seen, if n independent trials are made, we can define n independent random variables $\xi_1, \ldots, \xi_n$ on Ω^n by

$$\xi_i(\omega_1, \ldots, \omega_n) = \xi(\omega_i), \qquad 1 \leq i \leq n. \tag{1.4.1}$$

In the previous section we indicated roughly how the variables ξ_i can be used to make useful statistical predictions. In this section we demonstrate in somewhat greater detail how the theory might be applied. We shall discover that parameters associated with the unknown distribution of ξ may be approximated by parameters from a known sampling distribution generated by the values $\xi(\omega_i)$.

If the random variable ξ has the distribution function F, then so does each variable ξ_i. For a sequence of points $\omega_1, \omega_2, \ldots$ of Ω we shall refer to the numbers $t_1 = \xi(\omega_1), t_2 = \xi(\omega_2), \ldots$ as a sequence of *sample values* drawn from the distribution F.

Consider, for fixed n and each nonnegative integer k, the random variable

$$\alpha_k = \frac{1}{n} \sum_{i=1}^{n} \xi_i^k, \qquad k = 0, 1, 2, \ldots, \tag{1.4.2}$$

defined on the space Ω^n. For each sequence $\omega_1, \ldots, \omega_n$ of n points of Ω (corresponding to n trials of the experiment) the value of α_k at $(\omega_1, \ldots, \omega_n)$,

$$\alpha_k(\omega_1, \ldots, \omega_n) = \frac{1}{n} \sum_{i=1}^{n} t_i^k, \tag{1.4.3}$$

where $t_i = \xi(\omega_i)$, is called the k-th *sample moment* of the sample values $t_1, \ldots, t_n$. For each k, this number can be regarded as the k-th moment of a discrete distribution $F^*(t)$ defined by placing a mass of $1/n$ at each of the points t_i, $1 \leq i \leq n$. More precisely, $F^*(t)$ is a step function with jumps of height $1/n$ at each t_i:

$$F^*(t) = \frac{k}{n}, \qquad \text{where } k = \text{number of } t_i \leq t. \tag{1.4.4}$$

Then, clearly,

$$\alpha_k(\omega_1, \ldots, \omega_n) = \int_{-\infty}^{\infty} t^k \, dF^*(t) = \frac{1}{n} \sum_{i=1}^{n} t_i^k \tag{1.4.5}$$

is the k-th sample moment. The distribution $F^*(t)$ is sometimes called the *sampling distribution* associated with the sample values $t_1, \ldots, t_n$. The discrete distribution F^* provides an approximation of F in a sense made precise by the next theorem.

A fundamental tool in statistical analysis is the fact that under fairly general conditions a given characteristic of the sample (for example, the sample mean α_1) will converge, in some probability sense, to the corresponding characteristic of the parent population (population mean, $E[\xi]$) as the sample size (number of trials) tends to infinity. We shall say that a sequence of random variables $\xi_1, \xi_2, \ldots$ *converges in measure* to a random variable ξ if, for every $\epsilon > 0$,

$$\lim_{n \to \infty} P\{|\xi_n - \xi| \geq \epsilon\} = 0. \tag{1.4.6}$$

With this notion of convergence we may state the following general convergence theorem (see Ref. 3, p. 346) which gives us further information on the estimation problem we are trying to solve.

Theorem 1.4 Any rational function, or power of a rational function, of the random variable α_k converges in measure to the same function of the k-th moment of the parent distribution $F(t)$, provided all the moments concerned exist and are finite.

This result is important, for it tells us that, as we increase the sample size, our sample estimates will tend to get as close as we like to the numbers we are trying to estimate. Nevertheless, it gives us no control over the size of the error as a function of sample size. For this we again need a more precise analysis concerning the distribution of the random variable $\alpha_k - E[\xi^k]$ as n becomes large.

We first observe that the random variable

$$n\alpha_k = \sum_{i=1}^{n} \xi_i^k \tag{1.4.7}$$

is a sum of n mutually independent random variables all having the same distribution function and with expected value

$$E[\xi_i^k] = \tau_k = \int_{-\infty}^{\infty} t^k \, dF(t), \qquad 1 \leq i \leq n, \tag{1.4.8}$$

and variance

$$\sigma^2[\xi_i^k] = E[(\xi_i^k - \tau_k)^2] = \tau_{2k} - \tau_k^2, \qquad 1 \leq i \leq n. \tag{1.4.9}$$

It follows from the central limit theorem (Theorem 1.3) under the restriction $\sigma^2[\xi_i^k[< \infty$ that, as n tends to infinity, the distribution function of the random variable α_k is asymptotically normal $\left(\tau_k, \sqrt{(\tau_{2k} - \tau_k^2)/n}\right)$. In particular, when $k = 1$, we find as before that

$$\alpha_1 = \frac{1}{n} \sum_{i=1}^{n} \xi_i$$

is asymptotically normal with mean $\tau_1 = E[\xi]$ and standard deviation

$$\sqrt{\frac{\tau_2 - \tau_1^2}{n}} = \frac{\sigma}{\sqrt{n}},$$

where σ is the standard deviation of ξ. This result would enable us to make predictions about the error involved in using the quantity

$$\alpha_1(\omega_1, \ldots, \omega_n) = \frac{1}{n} \sum_{i=1}^{n} t_i$$

as an estimate of τ_1, provided we had some information available concerning σ. In practice, σ is usually unknown as well, and sample estimators must be used for it to give an approximation to the expected errors. For example, the random variable $\alpha_2 - \alpha_1^2$, whose value on a sequence $(\omega_1, \ldots, \omega_n)$ of trials is the sample variance

$$s^2 = (\alpha_2 - \alpha_1^2)(\omega_1, \ldots, \omega_n) = \frac{1}{n} \sum_{i=1}^{n} t_i^2 - \left(\frac{1}{n} \sum_{i=1}^{n} t_i\right)^2,$$

might be used to estimate σ^2 since, by Theorem 1.4, $\alpha_2 - \alpha_1^2$ converges in measure to σ^2. We say that $\alpha_2 - \alpha_1^2$ is a *consistent estimate* of σ^2 to express the fact that it converges in measure to σ^2. On the other hand, the expected value of $\alpha_2 - \alpha_1^2$ is not σ^2, but is $[(n-1)/n]\sigma^2$, as is easily shown. This means that, if we fix n and repeatedly draw samples $\omega_1, \ldots, \omega_n$ and calculate the sample variance s^2 each time, the average of all these values of s^2 will tend, not to σ^2, but to the smaller value $[(n-1)/n]\sigma^2$ instead. This may be remedied by replacing the estimate $\alpha_2 - \alpha_1^2$ by $[n/(n-1)](\alpha_2 - \alpha_1^2)$ so that

$$E\left[\frac{n}{n-1}(\alpha_2 - \alpha_1^2)\right] = \sigma^2. \tag{1.4.10}$$

An estimating random variable whose expected value is equal to the estimated quantity is called *unbiased*. Thus, $[n/(n-1)](\alpha_2 - \alpha_1^2)$ is both an unbiased and consistent estimate of the population variance σ^2 and is to be preferred, especially for small sample sizes.

There is one important distribution, Student's t-distribution, which arises in the analysis of normal variables. Let ξ be a normal (m, σ) random variable. Then, with α_k defined as in Eq. (1.4.3), the two random variables $\sqrt{n}(\alpha_1 - m)$ and $[n/(n-1)](\alpha_2 - \alpha_1^2)$ can be shown to be mutually independent, and $\sqrt{n}(\alpha_1 - m)$ is normal $(0, \sigma)$ while $[n/(n-1)](\alpha_2 - \alpha_1^2)$ is distributed as the arithmetic mean of $(n-1)$ squares of independent normal $(0, \sigma)$ variables. The ratio

$$t = \frac{\sqrt{n}(\alpha_1 - m)}{\sqrt{[n/(n-1)](\alpha_2 - \alpha_1^2)}} = \sqrt{n-1} \frac{\alpha_1 - m}{\sqrt{\alpha_2 - \alpha_1^2}} \tag{1.4.11}$$

has the density function

$$S_{n-1}(x) = \frac{1}{\sqrt{(n-1)\pi}} \frac{\Gamma\left(\dfrac{n}{2}\right)}{\Gamma\left(\dfrac{n-1}{2}\right)} \left(1 + \frac{x^2}{n-1}\right)^{-n/2}, \tag{1.4.12}$$

where

$$\Gamma(t) = \int_0^\infty x^{t-1} e^{-x} \, dx, \qquad t > 0, \tag{1.4.13}$$

is the Gamma function. The variable t is called *Student's ratio* and its distribution is called *Student's t-distribution with* $(n-1)$ *degrees of freedom.* The functions $S_{n-1}(x)$ have been tabulated (see, e.g., Ref. 3, p. 560) because of their frequent occurrence in applications. The reader will note that the Student t-distributions do not involve the population variance, σ^2. It is this fact, primarily, which accounts for their great utility. One can, with the aid of these distributions, develop a valuable test for the deviation of the estimating variable α_1 from the mean m of a normal distribution, a test which requires no knowledge of σ. In the following paragraph we show how such a test may be devised.

Let $\omega_1, \ldots, \omega_n$ represent n trials of the experiment and let $t_i = \xi(\omega_i)$ be the sample values, where ξ is a normal (m, σ) variable, m and σ being unknown. As before, let

$$\alpha_1(\omega_1, \ldots, \omega_n) = \frac{1}{n} \sum_{i=1}^n t_i \tag{1.4.14}$$

be the sample mean, and

$$s^2 = (\alpha_2 - \alpha_1^2)(\omega_1, \ldots, \omega_n) = \frac{1}{n} \sum_{i=1}^n t_i^2 - \left(\frac{1}{n} \sum_{i=1}^n t_i\right)^2 \tag{1.4.15}$$

be the sample variance. Since, by definition, the random variable

$$t = \sqrt{n-1} \, \frac{(\alpha_1 - m)}{\sqrt{\alpha_2 - \alpha_1^2}} \tag{1.4.16}$$

has the Student's t-distribution with $(n-1)$ degrees of freedom, we can write

$$P\{a < t \le b\} = \int_a^b S_{n-1}(x) \, dx \tag{1.4.17}$$

or

$$P\left\{\alpha_1 - \frac{b\sqrt{s^2}}{\sqrt{n-1}} \le m < \alpha_1 - \frac{a\sqrt{s^2}}{\sqrt{n-1}}\right\} = \int_a^b S_{n-1}(x) \, dx. \tag{1.4.18}$$

We have thus obtained a confidence interval for the population mean which is independent of σ in the following sense. Drawing repeated samples $\omega_1, \ldots, \omega_n$ of n from Ω and calculating for each sample the limits

$$\alpha = \frac{1}{n} \sum_{i=1}^n t_i - \frac{b\sqrt{s^2}}{\sqrt{n-1}}, \qquad \beta = \frac{1}{n} \sum_{i=1}^n t_i - \frac{a\sqrt{s^2}}{\sqrt{n-1}}, \tag{1.4.19}$$

the frequency of those intervals (α, β) which include the mean m will be approximately $\int_a^b S_{n-1}(x) \, dx$. The limits a and b may be chosen to give various values of this integral, called *confidence levels*, as close to unity as desired. It is clear, then, that (1.4.18) yields information concerning the deviation of the sample mean

$$\frac{1}{n}\sum_{i=1}^{n} t_i$$

from the true mean m for samples of size n.

One somewhat technical point about the above confidence interval theory seems worthy of further clarification. The quantities

$$\alpha_1 - \frac{b\sqrt{\alpha_2 - \alpha_1^2}}{\sqrt{n-1}} \quad \text{and} \quad \alpha_1 - \frac{\alpha\sqrt{\alpha_2 - \alpha_1^2}}{\sqrt{n-1}}$$

are random variables. Each fixed sample of size n gives rise to the numbers α, β and the sample mean

$$\frac{1}{n}\sum_{i=1}^{n} t_i.$$

Given these numbers, the mean m is either inside or outside the interval (α, β), but it is α and β which vary, not m. Thus, Eq. (1.4.18) should properly be interpreted as a statement about the frequency with which the interval (α, β) *contains m* (upon repeated drawing of n samples).

The preceding statistical analysis provides a valid means of obtaining confidence intervals for the mean of a normal population. When the population fails to be normal this analysis breaks down and, indeed, no general distribution-free methods are available for such cases. The assumption of normality thus becomes all-important and it is often useful to have a method available for testing this assumption.

There are many tests for normality; we shall mention some of them here and cite references for more complete discussions. We should point out that each test is capable of examining only one facet of the assumption of normality. In a sense, then, such a test can merely indicate a *tendency* toward normality. Furthermore, different tests will be sensitive in different ways to an absence of normality. For example, certain classical tests are rather insensitive to a departure from normality as long as the underlying distribution is symmetric.

Perhaps the best-known and most frequently applied tests for normality (and other distributional assumptions) are the chi-squared goodness of fit test, (Refs. 4, 5), the Kolmogoroff-Smirnoff test (Refs. 6–8), and the Cramér-Von Mises tests (Refs. 9–11). A new test procedure has recently been developed by Shapiro and Wilk (Ref. 12) which has two advantages. First, it is relatively easy to apply and, second, it is quite sensitive to deviations from normality. A complete description may be found in Ref. 12.

It is recommended that, where practical, some test, such as the Shapiro-Wilk test, be used to detect departures from normality in analyzing statistical data from Monte Carlo programs. When the departures from normality are sufficiently great, the confidence interval developed by the preceding discussion ceases to have validity and may not be relied upon. Of course, it is still the case that the sample mean provides the best estimate of the location of the population mean.

1.5 THE CONSTRUCTION OF SAMPLES

Basic to the use of Monte Carlo methods is the idea of drawing samples which are distributed according to a certain probability law. We begin our discussion of Monte Carlo methods proper with this topic.

Suppose that ξ is a random variable with density function $f(x)$. We want to draw samples from the density $f(x)$; that is, to construct a sequence of numbers $t_1, t_2, \ldots, -\infty \leq t_i \leq \infty$, so that

$$P\{a < t_i \leq b\} = \int_a^b f(x)\,dx \qquad \text{for } -\infty \leq a < b \leq \infty, \quad (1.5.1)$$

and

$$P\{a < t_{i_1}, \ldots, t_{i_n} \leq b\} = P\{a < t_{i_1} \leq b\} \cdots P\{a < t_{i_n} \leq b\}$$

$$= \left[\int_a^b f(x)\,dx \right]^n, \qquad \text{for } i_1, \ldots, i_n \text{ all different,}$$

$$(1.5.2)$$

where the left-hand side of (1.5.2) is the joint probability that $t_{i_1}, \ldots, t_{i_n}$ simultaneously satisfy the inequalities. The numbers $t_1, t_2, \ldots$ may be thought of as realizations of a sequence of random variables $\xi_1, \xi_2, \ldots$. Then Eq. (1.5.1) states that each ξ_i has density $f(x)$ and Eq. (1.5.2) states that $\xi_{i_1}, \ldots, \xi_{i_n}$ are mutually independent if the $i_1, \ldots, i_n$ are all different. There are various ways of constructing the numbers t_i; we begin with a description of perhaps the most basic method.

We first introduce the notion of a sequence of random numbers. The sequence $\rho_1, \rho_2, \ldots$ is a sequence of random numbers if the numbers ρ_i satisfy $0 \leq \rho_i \leq 1$ and if the ρ_i represent samples drawn independently from a uniform density on the unit interval $0 \leq x \leq 1$. That is,

$$P\{a < \rho_i \leq b\} = \int_a^b 1\,dx = b - a, \quad 0 \leq a < b \leq 1, \quad (1.5.3)$$

and

$$P\{a < \rho_{i_1}, \ldots, \rho_{i_n} \leq b\} = (b - a)^n, \qquad i_1, \ldots, i_n \text{ all different.} \quad (1.5.4)$$

Despite the flaw which is already evident in this discussion, we shall nevertheless postulate the existence of an indefinite supply of random numbers (and presently discuss the generation of such numbers at greater length). For the moment, then, the problem of drawing samples from an arbitrary density

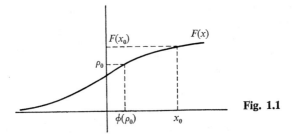

Fig. 1.1

$f(x)$ will be solved in terms of samples ρ_i from the uniform density

$$g(x) = \begin{cases} 1 & 0 \leq x \leq 1, \\ 0 & \text{otherwise.} \end{cases} \qquad (1.5.5)$$

Let $F(x)$ be the distribution function of ξ,

$$F(x) = P\{\xi \leq x\} = \int_{-\infty}^{x} f(t)\, dt \qquad (1.5.6)$$

and consider the equation

$$F(x) = \rho. \qquad (1.5.7)$$

As x varies over the interval $-\infty \leq x \leq \infty$, ρ ranges over the unit interval $0 \leq \rho \leq 1$. Now define a function $\phi(\rho)$ by

$$\phi(\rho) = \sup_{F(x) < \rho} x \qquad \text{(see Fig. 1.1).} \qquad (1.5.8)$$

Then it is clear that $\phi(F(x)) = x$ and $F(\phi(\rho)) = \rho$, so that F and ϕ are functional inverses to each other. From this it follows that

$$\phi(\rho) \leq x \qquad \text{if and only if} \qquad \rho \leq F(x). \qquad (1.5.9)$$

Now let $\rho_1, \rho_2, \ldots$ be a sequence of random numbers. Then the numbers

$$t_i = \phi(\rho_i) \qquad i = 1, 2, \ldots \qquad (1.5.10)$$

will be samples from the density $f(x)$, for

$$P\{t_i \leq x\} = P\{\phi(\rho_i) \leq x\} = P\{\rho_i \leq F(x)\} = F(x), \qquad (1.5.11)$$

the last equality following because the ρ_i are uniform. From (1.5.11) one may easily deduce (1.5.1) for any $a < b$ and the independence of the samples t_i similarly follows immediately from the independence of the ρ_i.

It may be useful, at this point, to restate our conclusions and to examine them more closely. We have found that if

a) ρ is uniformly distributed on the interval [0, 1] and
b) $\rho = \int_{-\infty}^{x} f(t)\, dt$,

then x has $f(x)$ as its density function. Through (b) we have imposed a functional relation between ρ and x and have shown, indirectly, that this relation between random variables implies a relation between their density functions. Actually we are dealing, here, with one special aspect of an important general principle. One can almost always deduce, from a relation between random variables, a simple relation between their distributions. Suppose, for example, that $G(\rho)$ and $F(x)$ are, respectively, distribution functions for ρ and x. Let $x = x(\rho)$ be a monotone increasing function of ρ. Draw N samples $\rho_1, \rho_2, \ldots, \rho_N$ from $G(\rho)$ and define

$$x_1 = x(\rho_1), \qquad x_2 = x(\rho_2), \ldots, \qquad x_N = x(\rho_N).$$

It follows from our definitions that the x_i may be regarded as samples drawn from $F(x)$. Assume that n of the ρ_i fall in the interval $[-\infty, \rho_c]$, so that n x_i's lie in the corresponding interval $[-\infty, x_c]$, $x_c = x(\rho_c)$. Clearly

$$\lim_{N \to \infty} \frac{n}{N} = G(\rho_c) = F(x_c). \tag{1.5.12}$$

Thus, for any corresponding ρ and x, $G(\rho) = F(x)$. If

$$G(\rho) = \int_{-\infty}^{\rho} g(t)\, dt \qquad \text{and} \qquad F(x) = \int_{-\infty}^{x} f(t)\, dt,$$

then

$$\frac{dG}{d\rho} = g(\rho) = \frac{dF}{dx}\frac{dx}{d\rho},$$

or

$$g(\rho) = f(x)\frac{dx}{d\rho}. \tag{1.5.13}$$

We shall use Eq. (1.5.13) whenever we have occasion to change from one random variable to another, related, random variable.

If it is known that ρ is uniform in $[0, 1]$, and that

$$F(x) = \int_{-\infty}^{x} f(t)\, dt,$$

then

$$G(\rho) = \rho = \int_{-\infty}^{x} f(t)\, dt, \tag{1.5.14}$$

$$x = \phi(\rho).$$

Again we see that the numbers $t_i = \phi(\rho_i)$ may be regarded as samples from the density $f(x)$.

To demonstrate the procedure described above, we consider the exponential density function

$$f(x) = \begin{cases} \Sigma e^{-\Sigma x}, & 0 \le x \le \infty, \quad \Sigma \text{ constant}, \\ 0, & x < 0. \end{cases} \tag{1.5.15}$$

It will be seen that $f(x)$ is the density function for distances travelled by neutrons between reactions in an infinite medium of material with total cross section Σ. The distribution function $F(x)$ is

$$F(x) = \begin{cases} 1 - e^{\Sigma x}, & x \geq 0, \\ 0, & x < 0. \end{cases} \tag{1.5.16}$$

To select samples from the density $f(x)$ one must solve the equation

$$\rho = F(x) = 1 - e^{-\Sigma x}, \qquad x \geq 0, \tag{1.5.17}$$

for x as a function of ρ. The solution is

$$x = \phi(\rho) = -\frac{1}{\Sigma} \ln (1 - \rho). \tag{1.5.18}$$

If random numbers ρ_i are substituted into the right-hand side of (1.5.18), the previous argument shows that the numbers

$$t_i = -\frac{1}{\Sigma} \ln (1 - \rho_i)$$

will be exponentially distributed in the interval $0 \leq x \leq \infty$. In practice, since the sequence $1 - \rho_1, 1 - \rho_2, \ldots$ is random if $\rho_1, \rho_2, \ldots$ is, the actual relationship is taken to be

$$t_i = -\frac{1}{\Sigma} \ln (\rho_i), \qquad i = 1, 2, \ldots. \tag{1.5.19}$$

Notice that the natural logarithm of a random number must be found for each sample generated. If a logarithm subroutine is available, it may of course be used, but the cost of each sample t_i will involve generating a random number ρ_i, finding its logarithm, and dividing by Σ. Sometimes, when great accuracy is not needed, a special-purpose logarithm subroutine may be written which sacrifices several decimal digits of accuracy in order to gain speed over a conventional routine. This may be accomplished, for example, by using fewer terms in a series or continued fraction expansion, or by using low-order rational function approximations to the logarithm. In this way the computing cost of generating a sample t_i may be reduced. An alternative method which is even more effective in lowering cost per sample is to store permanently a table of the logarithm (notice that in our example this need only be done on the interval $0 < x \leq 1$). This method, too, sacrifices accuracy for speed; one may use linear interpolation in this table to increase the accuracy in evaluating $\ln \rho_i$. Although this method has proven to be extremely useful in reducing computing time in Monte Carlo calculations, it does involve additional fast storage which may eventually tax the storage capacity of the computer. This is a common situation in Monte Carlo.

Because of the repetitive nature of the method, and because of the large number of samples required to obtain satisfactory statistics, it is extremely desirable to reduce the number of arithmetical and logical operations per sample to a minimum. This reduction may be aided by table storage, which places a burden on fast storage capacity and also sacrifices some accuracy, depending on the size of the table. In practice, a balance must be struck involving all of the factors speed, accuracy, and storage capacity. We shall later describe a compromise method, due to Marsaglia (Refs. 13–17), which allows the use of tables with no loss in accuracy.

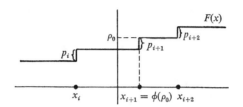

Fig. 1.2

The above-described method of drawing samples applies to any continuous distribution. A common problem in the application of Monte Carlo methods is that of sampling from a discrete distribution. As defined in Section 1.2, such a distribution is characterized by the property that it describes a countable number of mutually exclusive events, associated with the numbers x_i, the i-th event occurring with probability p_i, and with $\Sigma_i p_i = 1$. Let $F(x)$ be such a discrete distribution function; thus, we may write

$$F(x) = \sum_{x_i \le x} p_i. \tag{1.5.20}$$

Again we define a function $\phi(\rho)$ by

$$\phi(\rho) = \sup_{F(x_i) < \rho} x_i \qquad \text{(see Fig. 1.2)}. \tag{1.5.21}$$

From this we see that the range of choices

$$\sum_{i=0}^{j} p_i < \rho \le \sum_{i=0}^{j+1} p_i, \qquad p_0 = 0, \tag{1.5.22}$$

corresponds to the event $x_{j+1} = \phi(\rho)$, $j = 0, 1, \ldots$. The same argument used in the continuous case shows that if $\rho_1, \rho_2, \ldots$ is a sequence of random numbers, the quantities

$$x_k = \phi(\rho_k), \qquad k = 1, 2, \ldots \tag{1.5.23}$$

will constitute samples drawn from the discrete distribution $F(x)$. The prescription in the discrete case then takes the following form. Select a random number ρ and let j be chosen to satisfy (1.5.22). In other words, the inequality $F(x_j) < \rho \le F(x_{j+1})$ corresponds to the choice x_{j+1}.

We now give an example of drawing samples from a discrete distribution which frequently arises in our applications. If σ_t is the total microscopic cross section for a particular isotope, then σ_t is representable as a sum

$$\sigma_t = \sigma_1 + \sigma_2 + \cdots + \sigma_n, \tag{1.5.24}$$

where each σ_i denotes the cross section for a particular reaction, e.g. absorption, elastic scattering, fission, etc. The ratios $p_i = \sigma_i/\sigma_t$ then represent the probabilities for the various reactions i. Upon collision of a neutron with an atom of the isotope a random number ρ is generated and an integer j is found so that (1.5.22) is satisfied. This inequality is then identified with the reaction $j + 1$. If, for example, the reaction selected corresponds to elastic scattering, then other distributions would have to be sampled to determine the new direction and energy of the neutron after collision.

1.6 THE REJECTION METHOD

The method of Section 1.5 is a useful general method for sampling from a discrete or continuous distribution, but it may be computationally awkward for certain continuous distributions. This fact makes it desirable to consider other general methods. Another quite general method, called the rejection method, will be discussed in this section. The term rejection method is used because not all generated samples are used; some are rejected. Despite this waste, rejection techniques often result in simpler formulas, therefore less average cost per sample used, than the method of Section 1.5.

Let $f(x)$ be a bounded density function and assume f vanishes outside the interval (a, b). We wish to construct samples from the density $f(x)$. Let $M = \sup_{a \leq x \leq b} f(x)$ and define

$$f_1(x) = f(x)/M \tag{1.6.1}$$

so that $0 \leq f_1(x) \leq 1$ for $a \leq x \leq b$. Generate a pair (ρ_1, ρ_2) of random numbers and interpret $(a + \rho_1(b - a), \rho_2)$ as a point in the rectangle with base $(b - a)$ and height 1. If this point falls below the graph of $f_1(x)$, i.e. if $\rho_2 < f_1(a + \rho_1(b - a))$, then $t = a + \rho_1(b - a)$ is accepted as a sample from $f(x)$; if not, the process is repeated until a sample has been determined successfully.

This technique yields sample values t which are distributed according to the conditional density of t given that $\rho_2 < f_1(t)$. But the joint density of t and ρ_2, $g(t, \rho_2)$, is given by

$$g(t, \rho_2) = \begin{cases} \dfrac{1}{b - a} & \text{if } a \leq t \leq b, \quad 0 \leq \rho_2 \leq 1, \\ 0 & \text{elsewhere,} \end{cases} \tag{1.6.2}$$

since t and ρ_2 are independent and uniform. Therefore, the conditional density in question is (see Eq. 1.2.31)

$$g_1(t \mid \rho_2 < f_1(t)) = \frac{\left(\dfrac{1}{b-a}\right) \int_0^{f_1(t)} d\rho_2}{\left(\dfrac{1}{b-a}\right) \int_a^b dt \int_0^{f_1(t)} d\rho_2}$$

$$= \frac{f_1(t)}{\int_a^b f_1(t)\, dt}$$

$$= f(t), \tag{1.6.3}$$

so that the selected samples have the correct density.

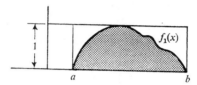

Fig. 1.3

As we noted earlier, the rejection technique suffers from the defect that not all of the random pairs (ρ_1, ρ_2) result in a sample t drawn from $f(x)$. In fact, the efficiency of such a technique, as measured by the fraction of pairs (ρ_1, ρ_2) which are not rejected, is just the ratio of the area under the curve $f_1(x)$ to the area of the enclosing rectangle (see Fig. 1.3). This ratio is clearly $1/[M(b-a)]$, and since

$$1 = \int_a^b f(x)\, dx \le (b-a)M, \tag{1.6.4}$$

this efficiency is properly ≤ 1. Let $E = 1/[M(b-a)]$ denote the efficiency. It is clear, then, that in practical comparisons between the rejection method and the inversion method presented in Section 1.5, the number of arithmetical and logical operations per sample for the inversion method should be compared with the product of $1/E$ and the number of operations for the rejection method, since they each represent the average work required per sample.

A further point worth mentioning is that, in order to use the rejection method, it is necessary to find $M = \sup_{a \le x \le b} f(x)$, or at least an upper bound for $f(x)$. If only a weak upper bound for $f(x)$ can be found, the efficiency of the rejection method will suffer correspondingly.

The following example illustrates an application of the simple type of rejection technique just described. Let the density function to be sampled be

$$f(x) = \begin{cases} \dfrac{4}{\pi(1 + x^2)}, & 0 \le x \le 1, \\[2mm] 0 & \text{elsewhere.} \end{cases} \tag{1.6.5}$$

Then $M = \sup_{0 \leq x \leq 1} f(x) = 4/\pi$. The distribution function corresponding to $f(x)$ is

$$F(x) = \begin{cases} 0, & x < 0, \\ \dfrac{4}{\pi} \displaystyle\int_0^x \dfrac{dt}{1 + t^2} = \dfrac{4}{\pi} \tan^{-1} x, & 0 \leq x \leq 1, \\ 1, & x > 1 \end{cases} \quad (1.6.6)$$

and is useful in Monte Carlo calculations for finding the tangent of an angle uniformly distributed between 0 and $\pi/2$. The rejection method consists of generating a pair (ρ_1, ρ_2) of random numbers and accepting ρ_1 as the sample value in case $\rho_2 < 1/(1 + \rho_1^2)$, or $\rho_2(1 + \rho_1^2) < 1$. The efficiency of the method is $\pi/4$ or about 79%, which is reasonably high for a rejection method.

The preceding method can be generalized in a variety of ways. Assume, for example, that the density function $f(x)$ can be factored:

$$f(x) = \frac{g(x)h(x)}{\int_{-\infty}^{\infty} g(x)h(x)\, dx}. \quad (1.6.7)$$

Here $g(x)$ is a density function and $h(x) \geq 0$. We choose a sample t from the density $g(x)$ and generate a random number ρ. If $\rho < h(t)$, the sample t is accepted. As before, the conditional density of t given that $\rho < h(t)$ is

$$p\big(t \mid \rho < h(t)\big) = \frac{g(t) \int_0^{h(t)} d\rho}{\int_{-\infty}^{\infty} g(t) [\int_0^{h(t)} d\rho]\, dt} = \frac{g(t)h(t)}{\int_{-\infty}^{\infty} g(t)h(t)\, dt}, \quad (1.6.8)$$

so that, by (1.6.7), $p\big(t \mid \rho < h(t)\big) = f(t)$. This method consists of a combination of techniques. The density $g(x)$ is sampled by whatever means are available and convenient, and rejection is used only on the function $h(x)$. This means that the restriction that $f(x)$ be bounded can be removed, as long as the function $f(x)/g(x)$ is bounded.

1.7 MULTI-DIMENSIONAL GENERALIZATIONS

In the previous sections we have described two general methods for constructing samples from one-dimensional density functions—the inversion method and the rejection technique. Before discussing more specialized techniques, we want to generalize the preceding descriptions to multi-dimensional density functions.

Let $f(x_1, \ldots, x_n)$ be a density function in n real variables. We want to draw samples from the density f; that is, to construct a sequence of n-tuples $(t_1^{(1)}, \ldots, t_n^{(1)})$, $(t_1^{(2)}, \ldots, t_n^{(2)}) \ldots$ with $-\infty \leq t_i^{(j)} \leq \infty$ for $1 \leq i \leq n, j = 1, 2, \ldots$ such that

$$P\{(t_1^{(j)}, \ldots, t_n^{(j)}) \in B_n\} = \int_{B_n} \cdots \int f(x_1, \ldots, x_n)\, dx_1 \cdots dx_n,$$
$$j = 1, 2, \ldots \quad (1.7.1)$$

and

$$P\{(t_1^{(j_1)}, \ldots, t_n^{(j_1)}), \ldots, (t_1^{(j_N)}, \ldots, t_n^{(j_N)}) \in B_n\}$$
$$= \left[\int_{B_N} \cdots \int f(x_1, \ldots, x_n) \, dx_1 \cdots dx_n \right]^N$$
$$\text{for } j_1, \ldots, j_n \text{ all different.} \quad (1.7.2)$$

In (1.7.1), the notation $(t_1^{(j)}, \ldots, t_n^{(j)}) \in B_n$ means the point $(t_1^{(j)}, \ldots, t_n^{(j)})$ belongs to the set B_n in n dimensions, where B_n is any set for which the integral on the right exists. As in the one-dimensional case, Eq. (1.7.1) expresses the fact that the samples are distributed with density f and Eq. (1.7.2) expresses the fact that the samples are mutually independent.

The multi-dimensional problem will be solved by reducing it to a series of one-dimensional sampling problems. Let $F(x_1, \ldots, x_n)$ be the distribution function associated with $f(x_1, \ldots, x_n)$,

$$F(x_1, \ldots, x_n) = \int_{-\infty}^{x_n} \cdots \int_{-\infty}^{x_1} f(y_1, \ldots, y_n) \, dy_1 \cdots dy_n. \quad (1.7.3)$$

Then $F(x_1, \ldots, x_n) = P\{(t_1, \ldots, t_n) \in R_n\}$, where $(t_1, \ldots, t_n)$ is any sample drawn from the density f and R_n is the semi-infinite rectangle in n dimensions defined by the inequalities $-\infty \leq y_1 \leq x_1, -\infty \leq y_2 \leq x_2, \ldots, -\infty \leq y_n \leq x_n$.

Let

$$f_1(x_1) = \int_{-\infty}^{\infty} \cdots \int_{-\infty}^{\infty} f(x_1, y_2, \ldots, y_n) \, dy_2 \cdots dy_n,$$

$$f_2(x_2 \mid x_1) = \frac{\int_{-\infty}^{\infty} \cdots \int_{-\infty}^{\infty} f(x_1, x_2, y_3, \ldots, y_n) \, dy_3 \cdots dy_n}{f_1(x_1)},$$

$$\vdots \qquad (1.7.4)$$

$$f_n(x_n \mid x_1, \ldots, x_{n-1}) = \frac{f(x_1, \ldots, x_n)}{\int_{-\infty}^{\infty} f(x_1, \ldots, x_n) \, dx_n},$$

so that each f_i, $1 \leq i \leq n$ is a one-dimensional density function in the variable x_i. The function $f_1(x_1)$ is what we have called the marginal density of x_1 (see Eq. 1.2.24), while for $i > 1$, $f_i(x_i \mid x_1, \ldots, x_{i-1})$ is the conditional density of x_i given $x_1, \ldots, x_{i-1}$. We observe that

$$f(x_1, \ldots, x_n) = \frac{\partial^n F(x_1, \ldots, x_n)}{\partial x_1 \partial x_2 \cdots \partial x_n}, \qquad f_1(x_1) \equiv \frac{\partial F(x_1, \infty, \ldots, \infty)}{\partial x_1}$$

$$\vdots \qquad (1.7.5)$$

$$f_i(x_i \mid x_1, \ldots, x_{i-1}) = \frac{\partial^i F(x_1, \ldots, x_i, \infty, \ldots, \infty)}{\partial x_1 \cdots \partial x_i}.$$

Consider the sequence of equations

$$F_1(x_1) = \int_{-\infty}^{x_1} f_1(y_1)\, dy_1 = \rho_1,$$

$$F_2(x_2 \mid x_1) = \int_{-\infty}^{x_2} f_2(y_2 \mid x_1)\, dy_2 = \rho_2,$$

$$\cdot$$
$$\cdot \qquad (1.7.6)$$
$$\cdot$$

$$F_n(x_n \mid x_1, \ldots, x_{n-1}) = \int_{-\infty}^{x_n} f_n(y_n \mid x_1, \ldots, x_{n-1})\, dy_n = \rho_n,$$

where $(\rho_1, \ldots, \rho_n)$ is an n-tuple of mutually independent random numbers. Each $F_i(x_i \mid x_1, \ldots, x_{i-1})$ is recognized as the conditional distribution function of x_i given $x_1, \ldots, x_{i-1}$. If functions

$$\phi_1(\rho_1), \phi_2(\rho_2 \mid x_1), \ldots, \phi_n(\rho_n \mid x_1, \ldots, x_{n-1})$$

are defined by

$$\phi_1(\rho_1) = \sup_{F_1(x_1) < \rho_1} x_1,$$
$$\phi_2(\rho_2 \mid x_1) = \sup_{F_2(x_2 \mid x_1) < \rho_2} x_2,$$
$$\cdot \qquad (1.7.7)$$
$$\cdot$$
$$\cdot$$
$$\phi_n(\rho_n \mid x_1, \ldots, x_{n-1}) = \sup_{F_n(x_n \mid x_1, \cdots, x_{n-1}) < \rho_n} x_n,$$

then, as in the one-dimensional case, it follows that $\phi_i(\rho_i \mid x_1, \ldots, x_{i-1})$ and $F_i(x_i \mid x_1, \ldots, x_{i-1})$ are functional inverses of each other. Therefore

$$\phi_i(\rho_i \mid x_1, \ldots, x_{i-1}) \leq x_i \quad \text{if and only if} \quad \rho_i \leq F_i(x_i \mid x_1, \ldots, x_{i-1}). \qquad (1.7.8)$$

We now show that if random numbers $\rho_1, \ldots, \rho_n$ are generated and numbers t_i defined by

$$t_1 = \phi_1(\rho_1),$$
$$t_2 = \phi_2(\rho_2 \mid t_1),$$
$$\cdot \qquad (1.7.9)$$
$$\cdot$$
$$\cdot$$
$$t_n = \phi_n(\rho_n \mid t_1, \ldots, t_{n-1}),$$

then the n-tuples $(t_1, \ldots, t_n)$ so generated will constitute samples from $f(x_1, \ldots, x_n)$. Referring to Eqs. (1.7.1) and (1.7.2), we shall consider only the case where B_n is the semi-infinite rectangle R_n. The proof for general sets B_n follows analogously. Now, with $(t_1, \ldots, t_n)$ defined by (1.7.9),

$$
\begin{aligned}
P\{(t_1, \ldots, t_n) \in R_n\} &= P\{-\infty \leq t_1 \leq x_1, \ldots, -\infty \leq t_n \leq x_n\} \\
&= P\{-\infty \leq \phi_1(\rho_1) \leq x_1, \ldots, \\
&\qquad -\infty \leq \phi_n(\rho_n \mid t_1, \ldots, t_{n-1}) \leq x_n\} \\
&= P\{0 \leq \rho_1 \leq F_1(x_1), \ldots, \\
&\qquad 0 \leq \rho_n \leq F_n(x_n \mid t_1, \ldots, t_{n-1})\} \\
&= \int_{-\infty}^{x_1} dF_1(t_1) \cdots \int_{-\infty}^{x_{n-1}} dF_{n-1}(t_{n-1} \mid t_1, \ldots, t_{n-2}) \\
&\qquad \times \int_{-\infty}^{x_n} dF_n(t_n \mid t_1, \ldots, t_{n-1}) \\
&= \int_{-\infty}^{x_1} f_1(t_1)\, dt_1 \cdots \int_{-\infty}^{x_{n-1}} f_{n-1}(t_{n-1} \mid t_1, \ldots, t_{n-2})\, dt_{n-1} \\
&\qquad \times \int_{-\infty}^{x_n} f_n(t_n \mid t_1, \ldots, t_{n-1})\, dt_n \\
&= \int_{-\infty}^{x_1} \cdots \int_{-\infty}^{x_n} f(t_1, \ldots, t_n)\, dt_1 \cdots dt_n \\
&= \int_{R_n} \cdots \int f(t_1, \ldots, t_n)\, dt_1 \cdots dt_n, \tag{1.7.10}
\end{aligned}
$$

as was to be shown. The independence of the samples follows easily from the independence of the random numbers, just as in the one-dimensional case.

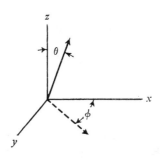

Fig. 1.4

As an example of this method, an example which occurs frequently in random walk problems, we consider the problem of drawing a vector at random from the unit sphere in three-dimensional space. Using polar coordinates (θ, ϕ), $0 \leq \theta \leq \pi$, $0 \leq \phi \leq 2\pi$ (see Fig. 1.4), the density function to be sampled is

$$
f(\theta, \phi) = \frac{\sin \theta\, d\theta\, d\phi}{4\pi}, \qquad 0 \leq \theta \leq \pi, \qquad 0 \leq \phi \leq 2\pi. \tag{1.7.11}
$$

Since the density f is a product, the two random variables are independent and the procedure simplifies to the following: let (ρ_1, ρ_2) be a pair of random function numbers and solve the equations

$$F_1(\theta) = \frac{1}{2} \int_0^\theta \sin t \, dt = \rho_1,$$

$$F_2(\phi \mid \theta) = \frac{1}{2\pi} \int_0^\phi dt = \rho_2 \qquad (1.7.12)$$

for θ, ϕ; i.e.

$$\theta = \arccos (1 - 2\rho_1),$$

$$\phi = 2\pi\rho_2. \qquad (1.7.13)$$

Frequently, it is necessary to calculate not θ, ϕ but $\sin \theta$, $\cos \theta$, $\sin \phi$, $\cos \phi$, where θ, ϕ have joint density function (1.7.11). We shall presently show how this may be accomplished without the need for sine and cosine subroutines.

A useful multi-dimensional generalization of the rejection method may be described in general terms as follows. Let $(t_1, \ldots, t_n)$ be a sample chosen from $f(x_1, \ldots, x_n)$ and set $z = h(t_1, \ldots, t_n)$ provided $t_n < g(t_1, \ldots, t_{n-1})$; otherwise reject the sample and choose another. The distribution function $G(z)$ of the random variable z defined in this way is given by the expression

$$G(z) = \frac{\displaystyle\int \cdots \int_{\substack{h(x_1, \ldots, x_n) < z \\ x_n < g(x_1, \ldots, x_{n-1})}} f(x_1, \ldots, x_n) \, dx_1 \cdots dx_n}{\displaystyle\int \cdots \int_{x_n < g(x_n, \ldots, x_{n-1})} f(x_1, \ldots, x_n) \, dx_1 \cdots dx_n}, \qquad (1.7.14)$$

where the inequalities define certain regions of n-dimensional space over which the integrations are performed.

To illustrate this procedure, let ρ_1, ρ_2 be a pair of random numbers and set

$$z = \frac{\rho_1^2 - \rho_2^2}{\rho_1^2 + \rho_2^2}, \qquad \text{if} \qquad \rho_2 < \sqrt{1 - \rho_1^2}.$$

It is easy to see by geometric arguments that z defines the cosine of an angle uniformly distributed between 0 and π. We shall actually compute the distribution of z, $-1 \leq z \leq 1$, using (1.7.14). We have

$$f(x_1, x_2) = \begin{cases} 1 & \text{if } 0 \leq x_1, x_2 \leq 1, \\ 0 & \text{elsewhere,} \end{cases}$$

$$h(x_1, x_2) = \frac{x_1^2 - x_2^2}{x_1^2 + x_2^2}, \qquad (1.7.15)$$

$$g(x_1) = \sqrt{1 - x_1^2}.$$

Then

$$G(z) = \frac{\iint_{B_1} dx_1 \, dx_2}{\iint_{B_2} dx_1 \, dx_2} = \frac{\iint_{B_2} dx_1 \, dx_2}{\pi/4}, \qquad (1.7.16)$$

where

$$B_1 = \{(x_1, x_2) \mid h(x_1, x_2) < z, 0 \le x_1, x_2 \le 1\}$$

and

$$B_2 = \{(x_1, x_2) \mid x_2 < g(x_1)\}.$$

By making the change of variables $s = x_1/x_2$, $t = x_2$ we find that

$$
\begin{aligned}
\iint_{B_1} dx_1 \, dx_2 &= \int_0^{\sqrt{(1+z)/(1-z)}} ds \int_0^{\sqrt{1/(1+s^2)}} t \, dt \\
&= \frac{1}{2} \int_0^{\sqrt{(z+1)/(1-z)}} \frac{ds}{1 + s^2} \\
&= \tfrac{1}{2} \arctan \sqrt{\frac{z+1}{1-z}}.
\end{aligned}
$$

Therefore,

$$G(z) = \frac{2}{\pi} \arctan \sqrt{\frac{z+1}{1-z}}, \qquad -1 \le z \le 1, \qquad (1.7.17)$$

and so

$$G'(z) = \frac{1}{\pi \sqrt{1 - z^2}}, \qquad -1 \le z \le 1, \qquad (1.7.18)$$

which may easily be shown to be the density of the random variable

$$z = \cos \pi \rho, \qquad \rho \text{ random.} \qquad (1.7.19)$$

Finally, we set

$$\sin \phi = \frac{\pm 2\rho_1 \rho_2}{\rho_1^2 + \rho_2^2}, \qquad (1.7.20)$$

where $\cos \phi = (\rho_1^2 - \rho_2^2)/(\rho_1^2 + \rho_2^2)$, the sign being chosen at random. In this way we obtain the sine and cosine of an angle uniformly distributed between 0 and 2π. This method, which was first suggested by von Neumann (Ref. 18), is widely used in Monte Carlo calculations.

1.8 SPECIAL METHODS

For the most part, the methods we have described in the past three sections suffice to generate samples from an arbitrary density in an efficient manner. Occasionally, however, certain combinations of methods or special tricks may be used to speed up the process of drawing samples. Kahn (Ref. 19) has presented methods for sampling from a long list of functions frequently encountered in Monte Carlo problems and the interested reader is referred to his discussion. When standard methods fail, a great deal of ingenuity is sometimes required to devise satisfactory sampling techniques.

Marsaglia (Refs. 13–17) has invented a class of sampling methods which have the advantage of being both general and fast. Briefly, these methods may be described as follows. Let $F(x)$ be an arbitrary distribution and write

$$F(x) = pF_1(x) + (1 - p)F_2(x), \qquad p < 1, \qquad (1.8.1)$$

where p is close to one and $F_1(x)$ is a distribution which is easy to sample. In practice, Marsaglia chooses $F_1(x)$ to be a discrete approximation to $F(x)$ (taking advantage of the fact that table look-up can be made exceptionally fast in a computer). The prescription then is: with probability p sample from $F_1(x)$, and with probability $(1 - p)$ sample from $F_2(x)$. If p is close enough to one, the average time per sample is little more than the time required to sample $F_1(x)$, despite the fact that the correction term $F_2(x)$ may be difficult to sample. Marsaglia has applied this technique to the normal distribution, writing

$$F(x) = p_1F_1(x) + p_2F_2(x) + (1 - p_1 - p_2)F_3(x), \qquad (1.8.2)$$

where $F_1(x)$ is a discrete approximation, $F_2(x)$ has a nearly linear density function, and $F_3(x)$ is the correction term. The result is a method which requires only enough storage for the table of $F_1(x)$, plus a few additional constants, and is only slightly slower per sample, on the average, than sampling from $F_1(x)$ alone. It has the advantage of being completely accurate† and appears to be potentially very useful for applications in a digital computer.

Other methods of generating samples from various density functions are based on the observation that it is sometimes possible to transform certain random variables by functions which give the required density.

An example of such transformations let

$$z = \max (\rho_1, \ldots, \rho_n), \qquad (1.8.3)$$

where $\rho_1, \ldots, \rho_n$ are n independent random numbers. Then one shows easily by induction that the density function of z is

$$f(z) = \begin{cases} nz^{n-1}, & 0 \le z \le 1, \\ 0 & \text{elsewhere.} \end{cases} \qquad (1.8.4)$$

---

† If $1 - p_1 - p_2$ is sufficiently small, then, *in practice*, the correction term $F_3(x)$ may never be sampled and Marsaglia's technique is effectively equivalent to using $p_1F_1 + p_2F_2$ as an approximation for $F(x)$. *In principle*, however, the method is completely accurate.

Since the distribution function of z is just

$$F(z) = z^n, \tag{1.8.5}$$

this example is useful in drawing samples from a polynomial density of the form

$$p(z) = \sum_{i=0}^{n} (i + 1)a_i z^i, \quad 0 \le z \le 1, \quad \text{where } a_i \ge 0 \text{ and } \sum_{i=0}^{n} a_i = 1. \tag{1.8.6}$$

The procedure is to let $z = \max(\rho_1, \ldots, \rho_i)$ with probability a_i: i.e. the numbers a_i are used to select a monomial term and then Eq. (1.8.3) is applied to this monomial.

Another useful transformation arises in the following way. Suppose it is desired to draw samples from a truncated exponential density

$$f(x) = \frac{e^{-x}}{1 - e^{-x_M}}, \quad 0 \le x \le x_M. \tag{1.8.7}$$

This density is useful in restricting particle collisions to take place at distances $d \le x_M$ from the point of origin. The ordinary inversion method may be used by calculating the distribution function

$$F(x) = \frac{1}{1 - e^{-x_M}} \int_0^x e^{-t}\, dt$$

$$= \frac{1 - e^{-x}}{1 - e^{-x_M}}, \quad 0 \le x \le x_M. \tag{1.8.8}$$

Then if ρ is a random number, the equation

$$\rho = F(x) \tag{1.8.9}$$

may be solved for x, yielding

$$x = -\ln\,[1 - \rho(1 - e^{-x_M})]. \tag{1.8.10}$$

This equation involves both a logarithm and an exponential, and may, therefore, result in a costly routine for generating samples. We shall discuss a trick, attributed to Coveyou by Leimdörfer (Ref. 20), which is employed to eliminate the exponential. Let t be a sample from the full exponential density

$$\pi(x) = e^{-x}, \quad 0 \le x \le \infty, \tag{1.8.11}$$

and let $\{t/x_M\}$ denote the full remainder upon division of t by x_M. Then $\{t/x_M\}$ is a random variable whose density we shall show to be (1.8.7). Let

$h(y)$ be the density of this remainder. Then, clearly,

$$h(y) = \pi(y) + \pi(y + x_M) + \pi(y + 2x_M) + \cdots$$
$$= e^{-y} + e^{-(y+x_M)} + e^{-(y+2x_M)} + \cdots$$
$$= e^{-y}(1 + e^{-x_M} + e^{-2x_M} + \cdots)$$
$$= \frac{e^{-y}}{1 - e^{-x_M}}, \qquad 0 \leq y \leq x_M, \qquad (1.8.12)$$

which was to be shown.

1.9 RANDOM NUMBER GENERATION

All of our sampling methods have been based on the assumption of an indefinite supply of random numbers. Therefore, we now turn to a discussion of how such numbers may be generated in a computer. As we have stated, random numbers are numbers between 0 and 1 which represent samples drawn independently from the uniform density function

$$f(x) = \begin{cases} 1, & 0 \leq x \leq 1, \\ 0 & \text{otherwise.} \end{cases} \qquad (1.9.1)$$

The idea of using deterministic processes to generate random numbers appears to have been advanced first in about 1946. Prior to that time, various random devices were used to generate random numbers. Such numbers have been produced by making "random selections" from census tables, phone directories, and by machines of one sort or another. RAND's table of a million random digits (Ref. 21) was generated using essentially an electronic roulette wheel, and some thought was given to building physical instruments which could feed random digits directly into early high-speed computers to be used as needed in the calculation. Such devices, however, make repetition of a calculation impossible and such repetition is indispensable in checking for errors.

Almost all of the present methods for generating random numbers on digital computers are based on the following observation: if two many-digit numbers x and y are multiplied together, the middle digits of the product xy appear to behave unpredictably as functions of the digits of x and y. This observation gave rise to an early method for generating approximately random numbers, the mid-square technique. This method consists of a recursion formula whereby the $(n + 1)$-th random number is obtained from the n-th by squaring it and extracting the appropriate number of digits from the middle of the square. The idea is that, although the entire sequence of numbers is completely determined once the first number is specified, the numbers behave statistically as though they were drawn at random. That is, they are roughly uniform and roughly uncorrelated, at least sufficiently so that sampling errors made by using them are small. Such numbers are frequently referred to as *pseudo-random numbers*. The mid-square method of generating pseudo-random numbers has often been used in Monte Carlo calculations with various degrees of success reported in the literature.

In 1949 a variation of the mid-square method was suggested by Lehmer (Ref. 22). This algorithm, called the multiplicative congruential method, takes the form†

$$x_i \equiv ax_{i-1}(\bmod m), \tag{1.9.2}$$

where x_0 is a positive integer, a is a positive integer, and the modulus, m, is a positive and large integer. A sequence $\{x_i\}$ of nonnegative integers is formed by multiplying x_{i-1} by a and then reducing the product modulo m. The pseudo-random numbers themselves are generated by forming the sequence $\{x_i/m\}$. In practice, the modulus m is taken to be a high power of 2 on a binary computer and a high power of 10 on a decimal computer. This stratagem permits the divisions by m implicit in the algorithm to be avoided and increases the speed of execution of the algorithm.

The primary questions concerning the algorithm (1.9.2) are:

a) how many distinct numbers x_i are produced by it? (clearly no more than m), and

b) how well does the sequence $\{x_i/m\}$ behave statistically with respect to randomness?

The number of distinct x_i which are generated before repetition is called the *period* of the pseudo-random number sequence. The study of choices of a, x_0, and m which lead to sequences with large periods depends on number-theoretic considerations. Such studies have been made by a number of authors (Refs. 23–26). It turns out that, for the multiplicative congruential method, a period of length m cannot be guaranteed; nevertheless, a variety of choices of a, x_0, and m will ensure that the period is large, i.e. nearly m. Furthermore, it appears that if the sequences do not repeat too quickly, the numbers $\{x_i/m\}$ hold up quite well under a variety of tests for randomness. Various choices of a, x_0, and m which lead to useful sequences have been reported in the literature and summarized in Ref. 27.

A variation of the multiplicative congruential method, called the mixed congruential method, is defined by

$$x_i \equiv [ax_{i-1} + c](\bmod m), \tag{1.9.3}$$

where c is again a suitably chosen integer. The main effect of the addition of c is to ensure long periods in the least significant digits of x_i as well as in the most significant digits. Multiplicative methods suffer from the defect that the period decreases as one moves from the most significant to the least

---

† The notation $x \equiv y$ (mod m) is read "x is congruent to y modulo m" and means x is the remainder of y upon division by m, $0 \leq x < m$.

significant digits. Analysis shows that, with the mixed methods, it is easy to guarantee full period and the results of statistical testing have proved satisfactory in most cases. Some recent work (Refs. 28–30) has been concerned with the serial correlation between numbers of such pseudo-random sequences. Such studies are designed to enable one to choose sequences in which the numbers of the sequence are as uncorrelated as possible. This condition would, of course, be satisfied ideally in the case of random numbers but it is probably true that some correlation can be tolerated in many Monte Carlo applications. An excellent discussion of pseudo-random number generators, with an extensive bibliography, is given in Ref. 27.

Marsaglia (Ref. 31) has studied a class of linear transformations suggested by formula (1.9.3) and his study reveals interesting properties of such transformations. Suppose, for example, that x is a uniform random variable on $0 \leq x \leq 1$ and consider the linear transformation

$$T(x) = [nx + c](\text{mod } 1), \qquad (1.9.4)$$

where n is an integer and $0 \leq c < 1$. From the preceding discussion, one would expect the sequence x, $T(x)$, $T(T(x))$, ..., to behave much like a sequence of independent, uniform variables. Marsaglia points out that the point $(x, T(x))$ is uniformly distributed over the graph of the function T. Now, for n sufficiently large, the graph of T more and more nearly fills the unit square, so that, while $T(x)$ is completely determined by x, the pair $(x, T(x))$ serves very nicely in place of a pair of independent uniform variables. The reader may convince himself of these facts by inspecting the graph of T for increasing values of n. Marsaglia's interesting analysis also shows that the correlation between x and $T(x)$ is negligible for sufficiently large n and exhibits a pair of values for c for which this correlation will vanish for any n.

From this point on in the discussion we shall assume that an indefinite supply of pseudo-random numbers is available to us.

REFERENCES

1. P. R. HALMOS, *Measure Theory*, D. Van Nostrand Co., Inc., New York (1950).

2. M. LOÈVE, *Probability Theory*, D. Van Nostrand Co., Inc., New York (1950).

3. H. CRAMÉR, *Mathematical Methods of Statistics*, Princeton University Press, Princeton, N.Y. (1946).

4. J. NEYMAN, and E. S. PEARSON, "Further Notes on χ^2 Distribution," *Biometrika*, **22**, 298 (1930).

5. H. CHERNOFF, and E. L. LEHMANN, "The Use of Maximum Likelihood Estimates in χ^2 Tests of Goodness of Fit," *Ann. Math. Stat.*, **25**, 579 (1954).

6. A. N. KOLMOGOROFF, "Sulla Determinazione Empirica di une Legge di Distribuzione," *Giorn. dell' Inst. Ital. degli Attuari*, **4**, 83 (1933).

7. N. V. SMIRNOV, "Sur les Scorto de la Courbe de Distribution Empirique," *Rec. Math. (NS) [Mat. Sborn]*, **6**, 3 (1939).

8. N. V. SMIRNOV, "Table for Estimating the Goodness of Fit of Empirical Distributions," *Ann. Math. Stat.*, **19**, 279 (1948).

9. H. CRAMÉR, "On the Composition of Elementary Errors," *Skand. Aktuar.*, **11**, 141 (1928).

10. R. VON MISES, *Wahrscheinlichkeitsrechnung*, Dentiske, Leipzig and Wien (1931).

11. T. W. ANDERSON, and D. A. DARLING, "A Test for Goodness of Fit," *J. Am. Stat. Assoc.*, **49**, 765 (1954).

12. S. S. SHAPIRO, and M. B. WILK, "An Analysis of Variance Test for Normality (complete samples)," *Biometrika*, **52**, 591 (1965).

13. G. MARSAGLIA, "Expressing a Random Variable in Terms of Uniform Random Variables," *Ann. Math. Stat.*, **32**, 894 (1961).

14. G. MARSAGLIA, "Generating Exponential Random Variables," *Ann. Math. Stat.*, **32**, 899 (1961).

15. G. MARSAGLIA, "Remark on Generating a Random Variable Having a Nearly Linear Density Function," *Boeing Scientific Research Labs. Mathematical Note No. 242* (Seattle, 1961).

16. G. MARSAGLIA, M. D. MACLAREN, and T. A. BRAY, "A Fast Procedure for Generating Normal Random Variables," *Comm. ACM*, **7**, (1) 4 (January, 1964).

17. M. D. MACLAREN, G. MARSAGLIA, and T. A. BRAY, "A Fast Procedure for Generating Exponential Random Variables," *Comm. ACM*, **7**, (5) 298 (May, 1964).

18. J. VON NEUMANN, "Various Techniques Used in Connection with Random Digits," *Natl. Bur. Stand. (U.S.), Appl. Math. Series* 12, 36 (1951).

19. HERMAN KAHN, *Applications of Monte Carlo*, Rand Corp. AECU-3259 (April, 1954; Revised April, 1956).

20. MARTIN LEIMDÖRFER, "On the Transformation of the Transport Equation for Solving Deep Penetration Problems by the Monte Carlo Method," *Trans. Chalmers Univ. Technol., Gothenberg*, **286** (1964).

21. Rand Corp., *A Million Random Digits with 100,000 Normal Deviates*, The Free Press, Glencoe, Illinois (1955).

22. D. H. LEHMER, "Mathematical Methods in Large-Scale Computing Units," *Proc. 2nd Symp. on Large-Scale Digital Calculating Machinery*, pp. 141–146 (1949); *Ann. Comp. Lab. Harvard Univ.*, **26** (1951).

23. J. CERTAINE, "On a Sequence of Pseudo-Random Numbers of Maximal Length," *J. Assoc. Comp. Mach.*, **5**, 353 (1958).

24. OLGA TAUSSKY, and JOHN TODD, "Generation and Testing of Pseudo-Random Numbers," *Symposium on Monte Carlo Methods*, ed. H. A. Meyer, Wiley, New York (1956), pp. 15–28.

25. H. A. MEYER, L. S. GEPHART, and N. L. TASMUSSEN, "On the Generation and Testing of Random Digits," *Air Res. Dev. Command, WADC Tech. Rep.* 54–55, Wright-Patterson Air Force Base (1954).

26. E. S. PAGE, "Pseudo-Random Elements for Computers," *Appl. Stat.*, **8**, 124 (1959).

27. T. E. HULL, and A. R. DOBELL, "Random Number Generators," *SIAM Rev.*, **4**, 230 (1962).

28. R. R. COVEYOU, "Serial Correlation in the Generation of Pseudo-Random Numbers," *J. Assoc. Comp. Mach.*, **7**, 72 (1960).

29. M. GREENBERGER, "Notes on a New Pseudo-Random Number Generator," *J. Assoc. Comp. Mach.*, **8**, 163 (1961).

30. M. GREENBERGER, "An A Priori Determination of Serial Correlation in Computer Generated Random Numbers," *Math. Comp.*, **15**, 383 (1961); Corrigenda, *Math. Comp.*, **16**, 126 (1962).

31. GEORGE MARSAGLIA, "Random Variables and Computers," *Trans. Third Prague Conf. Information Theory, Statist. Decision Functions, Random Processes* (*Liblice, 1962*), pp. 499–512. Publ. House Czech. Acad. Sci., Prague (1964).

2

Discrete and Continuous Random Walk Processes

2.1 INTRODUCTION

In Chapter 2 we turn our attention from sampling theory and its underlying probability foundations to the special techniques of Monte Carlo. In expository work on Monte Carlo it is customary to begin with a discussion of the problem of quadrature, the problem of evaluating integrals. We choose, however, to part with tradition and start with another problem instead. In the opening sections of this chapter we shall deal, not with integrals, but with sets of linear algebraic equations. Many of the principles of Monte Carlo will be illustrated by showing how Monte Carlo may be used to solve such equations. This approach has two features which we find attractive. First, it is relatively easy to move in small steps from the solution of matrix equations, to the solution of certain simple integral equations, to the solution of the transport equation. In fact, using the equation for neutron transport in an infinite medium as a model, ideas from transport theory can be injected quite naturally into a Monte Carlo treatment of matrix equations. Second, the solution in Chapter 5 of the thermal multigroup transport equations involves a mixture of matrix and integral equation techniques. This results from combining a discrete treatment of the energy variable with a continuous treatment of spatial variables. Implicit use of our discussion of Monte Carlo methods applied to linear systems is made in the energy treatment of Chapter 5.

In Chapter 2 we restrict our attention for simplicity to non-multiplying, subcritical systems. The generalization to multiplying, subcritical systems will be treated for the continuous case in Chapter 3. As we shall see there, the generalization is not difficult. Still, by avoiding multiplication in the present chapter, many formulas are simplified, and we hope the exposition is made smoother as a result.

We should also point out that in Chapter 2, as well as in Chapter 3, we deal almost exclusively with the integral form of the transport equation. We do this because this form of the equation seems best suited to the derivation of general results about estimators. We believe that this is so because the integral equation (and its discrete counterpart, the matrix equation) lends itself so naturally to a probability interpretation, an interpretation which we try to reveal in the first five sections of Chapter 2. Nevertheless, the integro-differential form is customarily used in discussions in which transport theory *per se* is stressed. In Chapters 4–6 we shall find topics in transport theory treated as an essential part of the applications of the Monte Carlo theory developed in Chapters 1–3. Therefore, in accord with custom, we turn our attention in these latter chapters from a treatment based on integral equations to one based on the integro-differential equation. In Section 2.4 we shall derive some relations between various integral transport equations and the integro-differential form which we hope will help make contacts between these two somewhat different points of view.

2.2 DISCRETE RANDOM WALKS

As we have stated earlier, the Monte Carlo processes in which we are interested will simulate the transport of particles (neutrons, gamma-rays, etc.) from point to point in phase space. In general, the phase space will be a six-dimensional continuum and the transport equation gives a mathematical description of the processes we shall study. It is instructive, however, to consider the transport of particles in a discrete, finite phase space, from which we are led to a matrix equation as a mathematical description of the process. We shall use this simpler model to develop many of the ideas which will later be applied to more complex problems.

We shall begin by developing an equation for a discrete collision density. We postulate a phase space consisting of N possible states which we label with the indices $1, \ldots, N$. A particle random walk on such a phase space is characterized by a first collision in some state i_1, subsequent transmission through a sequence of states $i_2, i_3, \ldots$ and, finally, with probability 1, termination in some state i_k. A *discrete random walk process*† will therefore be completely specified by a set of first collision probabilities, $p_i^1 =$ probability of first collision in state i; a set of transmission probabilities, $p_{i,j} =$ probability of transition from state j to state i;‡ and a set of death probabilities, $p_i =$ probability of termination in state i. If we let $\alpha = (i_1, \ldots, i_{k-1}, i_k)$ denote a typical random walk, k being the number of collisions made to

---

† It will be the case throughout this book that the term random walk process signifies both a set of probability functions and the process of sampling such functions in the execution of a Monte Carlo simulation. We trust that no confusion will result from this dual use of the term.

‡ This "backwards" notation is based on conventions adopted in discussing integral equations (see Eq. 2.4.6).

termination, then

$$p_i^1 = P(i_1 = i),$$

$$p_{i,j} = P(i_{n+1} = i \mid i_n = j \text{ and } k > n), \qquad (2.2.1)$$

$$p_i = P(k = n \mid i_n = i).$$

We require that $p_i > 0$, $p_{i,j} \geq 0$,

$$\sum_{i=1}^{N} p_i^1 = 1, \qquad\qquad \text{and}$$

$$\qquad\qquad\qquad\qquad\qquad\qquad\qquad\qquad (2.2.2)$$

$$\sum_{i=1}^{N} p_{i,j} = 1 - p_j \leq 1 \qquad \text{for all } j.$$

The latter condition is a consequence of the exclusion of multiplication. We shall say that the random walk α has made collisions at $i_1, \ldots, i_k$.

With this intuitive notion of collision, let P_j^n denote the probability of making collision n at index j:

$$P_j^n = P(i_n = j \mid k > n - 1) \qquad n \geq 2. \qquad (2.2.3)$$

We have

$$P_j^1 = P(i_1 = j) = p_j^1, \qquad\qquad (2.2.4)$$

and it is clear that a recursion formula for P_j^n is

$$P_j^n = \sum_{k=1}^{N} p_{j,k} P_k^{n-1}, \qquad n \geq 2. \qquad (2.2.5)$$

Equation (2.2.5) expresses the fact that the probability that collision n takes place at j is the probability that the $(n - 1)$-th collision takes place at k, times the probability of transition from k to j, summed over all intermediate states k.

For each state j define a random variable X_j by $X_j = $ number of collisions made at j. Then X_j is a discrete random variable on the space of all random walks α, which may take on any positive integral value. It is not hard to see that the expected value of X_j, $E[X_j]$, may be calculated by

$$E[X_j] = 1 \cdot P_j^1 + 1 \cdot P_j^2 + 1 \cdot P_j^3 + \cdots$$

$$= \sum_{n=1}^{\infty} P_j^n. \qquad\qquad (2.2.6)$$

We shall call $P_j = E[X_j]$ the *discrete collision density* at j, for obvious reasons.

To calculate P_j, using Eq. (2.2.5),

$$\sum_{n=2}^{\infty} P_j^n = \sum_{n=2}^{\infty} \sum_{k=1}^{N} p_{j,k} P_k^{n-1}$$

$$= \sum_{k=1}^{N} p_{j,k} \sum_{n=2}^{\infty} P_k^{n-1}$$

$$= \sum_{k=1}^{N} p_{j,k} \sum_{n=1}^{\infty} P_k^n.$$

Therefore,

$$P_j = P_j^1 + \sum_{k=1}^{N} p_{j,k} P_k, \qquad 1 \leq j \leq N. \tag{2.2.7}$$

If we denote the vector $(P_1, \ldots, P_n)$ by $\mathbf{P}$ and the matrix with entries $p_{i,j}$ by K, then the system of Eqs. (2.2.7) may be written as a matrix equation

$$\mathbf{P} = \mathbf{P}^1 + K\mathbf{P}, \tag{2.2.8}$$

where

$$\mathbf{P}_1 = (P_1^1, \ldots, P_N^1).$$

The matrix equation just derived relates the discrete collision density $\mathbf{P}$ to a density of first collisions $\mathbf{P}^1$ and an operator K which describes the probability of direct transition from one state to another.

The solution of Eq. (2.2.8) is a vector $\mathbf{P}$ whose j-th component, P_j, may be described as the density of particles about to undergo collision in state j. Thus, the source term, $\mathbf{P}^1$, is interpreted as the density of particles about to undergo a first collision, particles which have already been transferred from their birth state to the state at which the first collision will be made. We need not be concerned here either with the birth process itself or the mechanism transferring particles from a birth state to a first collision state. In later sections dealing specifically with the transport equation, however, a distinction will have to be made between the density of particles arising from births and a density of first collisions.

As a realistic model of the above process we may use the multigroup neutron transport equations, specialized to an infinite medium. These may be derived in the following way.

We begin with the multigroup transport equations† for a (possibly) heterogeneous system without multiplication:

$$\boldsymbol{\omega} \cdot \nabla F^i(\mathbf{r}, \boldsymbol{\omega}) + \Sigma_t^i(\mathbf{r}) F^i(\mathbf{r}, \boldsymbol{\omega})$$

$$= \sum_{j=1}^{G} \iint d\boldsymbol{\omega}' \Sigma_t^j(\mathbf{r}) C^{ij}(\mathbf{r}, \boldsymbol{\omega} \cdot \boldsymbol{\omega}') F^j(\mathbf{r}, \boldsymbol{\omega}') + Q^i(\mathbf{r}, \boldsymbol{\omega}), \tag{2.2.9}$$

where i is the energy group index, $1 \leq i \leq G$; ∇ is the gradient vector

---

† The derivation of the multigroup transport equations from the general transport equation is indicated in Section 5.2.

operator; $\boldsymbol{\omega}$ is a unit direction vector; $\mathbf{r}$ is a spatial vector; $F^i(\mathbf{r}, \boldsymbol{\omega})$ is the vector flux† at $(\mathbf{r}, \boldsymbol{\omega})$ in the i-th group; $\Sigma_t^i(\mathbf{r})$ is the total macroscopic cross section in group i; $C^{ij}(\mathbf{r}, \boldsymbol{\omega} \cdot \boldsymbol{\omega}')$ is the scattering kernel for all elements from group j to group i and from direction $\boldsymbol{\omega}'$ to $\boldsymbol{\omega}$; and $Q^i(\mathbf{r}, \boldsymbol{\omega})$ is the source at $(\mathbf{r}, \boldsymbol{\omega})$ in group i. In Eq. (2.2.9) the kernel C^{ij} is a composite scattering kernel which depends on all elements present in the medium.

In an infinite homogeneous medium containing a constant source the F^i are constant, so that $\nabla F^i = 0, 1 \leq i \leq G$. If, further, we assume that the source is isotropic, $Q^i(\boldsymbol{\omega}) = Q^i/4\pi$, the multigroup equations become

$$\Sigma_t^i F^i(\boldsymbol{\omega}) = \sum_{j=1}^{G} \iint d\boldsymbol{\omega}' \Sigma_t^j C^{ij}(\boldsymbol{\omega} \cdot \boldsymbol{\omega}') F^j(\omega) + \frac{Q^i}{4\pi}, \quad 1 \leq i \leq G. \quad (2.2.10)$$

Let

$$\phi^i \equiv \iint d\boldsymbol{\omega} F^i(\boldsymbol{\omega})$$

be the scalar flux in group i and integrate Eq. (2.2.10) over all directions. The result is

$$\Sigma_t^i \phi^i = \sum_{j=1}^{G} \Sigma_t^i C^{ij} \phi^i + Q^i, \quad 1 \leq i \leq G, \quad (2.2.11)$$

the infinite medium equations for the scalar flux ϕ^i. Here

$$C^{ij} \equiv \iint C^{ij}(\boldsymbol{\omega} \cdot \boldsymbol{\omega}') \, d\omega' = \frac{\Sigma_s^{ij}}{\Sigma_t^i}$$

and the matrix Σ_s^{ij} is the transition matrix from energy group j to energy group i. Thus,‡

$$\sum_{i=1}^{G} \Sigma_s^{ij} = \Sigma_s^j \leq \Sigma_t^j, \quad 1 \leq j \leq G,$$

where Σ_s^j is the macroscopic scattering cross section in group j and where Σ_t^i is the total macroscopic cross section in group i.

Let

$$\theta^i = \Sigma_t^i \phi^i$$

so that θ^i is the macroscopic (scalar) collision density in group i. Then (2.2.11) becomes

$$\theta^i = \sum_{j=1}^{G} \frac{\Sigma_s^{ij}}{\Sigma_t^j} \theta^j + Q^i, \quad 1 \leq i \leq G \quad (2.2.12)$$

---

† The flux F^i and macroscopic collision density ψ^i are related by $\psi^i = \Sigma_t^i F^i$.
‡ This relation between the scattering cross section Σ_s^j and matrix elements Σ_s^{ij} really depends on the manner of deriving multigroup equations (cf. discussion in Chapter 5, where these multigroup equations are derived).

under the assumption that $\Sigma_t^j \neq 0$ for all j.† Thus, we have the matrix equation

$$\theta = K\theta + Q, \tag{2.2.13}$$

where

$$\theta = \begin{bmatrix} \theta^1 \\ \cdot \\ \cdot \\ \cdot \\ \theta^G \end{bmatrix}, \quad Q = \begin{bmatrix} Q^1 \\ \cdot \\ \cdot \\ \cdot \\ Q^G \end{bmatrix}, \quad K = \begin{bmatrix} \dfrac{\Sigma_s^{11}}{\Sigma_t^1} & \dfrac{\Sigma_s^{12}}{\Sigma_t^2} & \cdots & \dfrac{\Sigma_s^{1G}}{\Sigma_t^G} \\ \cdot & & & \cdot \\ \cdot & & & \cdot \\ \cdot & & & \cdot \\ \dfrac{\Sigma_s^{G1}}{\Sigma_t^1} & \dfrac{\Sigma_s^{G2}}{\Sigma_t^2} & \cdots & \dfrac{\Sigma_s^{GG}}{\Sigma_t^G} \end{bmatrix}.$$

We note that on physical grounds, $Q \geq 0$, $K \geq 0$, and $\theta \geq 0$.‡
Now define

$$p_{ij} = \frac{\Sigma_s^{ij}}{\Sigma_t^j}$$

so that $0 \leq p_{ij} \leq 1$ and

$$\sum_{i=1}^{G} p_{ij} = \frac{1}{\Sigma_t^j} \sum_{i=1}^{G} \Sigma_s^{ij} = \frac{\Sigma_s^j}{\Sigma_t^j}.$$

Let

$$p_j = 1 - \frac{\Sigma_a^j}{\Sigma_t^j},$$

where $\Sigma_a^j = \Sigma_t^j - \Sigma_s^j$ is the macroscopic absorption cross section in group j. Further, define

$$p_i^1 = \frac{Q^i}{\sum_{i=1}^{G} Q_i}$$

so that $0 \leq p_i^1 \leq 1$ and $\sum_{i=1}^{G} p_i^1 = 1$. With these definitions, requirements (2.2.1) and (2.2.2) for a discrete random walk process have been satisfied.
Then, dividing Eq. (2.2.13) by the normalizing factor

$$\frac{1}{\sum_{i=1}^{G} Q^i},$$

we have

$$\frac{1}{\sum_{i=1}^{G} Q_i} \theta = K \left(\frac{1}{\sum_{i=1}^{G} Q^i} \theta \right) + \frac{1}{\sum_{i=1}^{G} Q^i} Q.$$

---

† If $\Sigma_t^j = 0$ for some j, the order of the linear system may be reduced by decoupling group j from the remaining $(G - 1)$ groups.
‡ We write $x \geq 0$ if $x_i \geq 0$ for all i and $A \geq 0$ if $a_{ij} \geq 0$ for all i and j, where x is a vector and A is a matrix.

Finally, defining

$$\mathbf{P} = \frac{1}{\sum_{i=1}^{G} Q^i} \, \boldsymbol{\theta}, \qquad \mathbf{P^1} = \frac{1}{\sum_{i=1}^{G} Q^i} \, \mathbf{Q},$$

we have

$$\mathbf{P} = \mathbf{P^1} + K\mathbf{P}. \qquad (2.2.14)$$

We have succeeded in identifying the multigroup transport process in an infinite medium as a discrete random walk process of the type described earlier. In particular, $\mathbf{P}$ is the normalized vector collision density whose j-th component, P_j, represents the expected number of collisions in group j per unit source particle.

We notice that, here, the source vector $\mathbf{P^1}$ may be identified either as the density of births or the density of first collisions; they are the same, since there is no spatial dependence and a particle born in energy group i will make its first collision there.

2.3 SOLUTIONS OF LINEAR SYSTEMS

In this section we shall discuss, in the context of discrete random walk processes, methods for estimating the solution of a system of linear algebraic equations. Some of the methods we discuss were among the first Monte Carlo applications, but our presentation of them differs somewhat from the usual. This is because we have aimed the presentation at the solution of the multigroup infinite medium equations. The result is that, for the most part, estimators are developed which generalize to the case of ultimate interest, namely, the integral form of the Boltzmann transport equation.

Let

$$\mathbf{x} = H\mathbf{x} + \mathbf{a} \qquad (2.3.1)$$

be a general matrix equation, with H a real $N \times N$ matrix and $\mathbf{a}$ a known vector. On the space of all real N vectors define the usual inner product by

$$\langle \mathbf{x}, \mathbf{y} \rangle = \sum_{i=1}^{N} x_i y_i. \qquad (2.3.2)$$

Then, if we denote the transpose of a matrix A by A', $(A')_{ij} = A_{ji}$, we have

$$\langle A'\mathbf{x}, \mathbf{y} \rangle = \langle \mathbf{x}, A\mathbf{y} \rangle \qquad (2.3.3)$$

for all vectors $\mathbf{x}$, $\mathbf{y}$. Let

$$\mathbf{y} = H'\mathbf{y} + \mathbf{b} \qquad (2.3.4)$$

be the equation adjoint† to Eq. (2.3.1), where $\mathbf{b}$ is arbitrary. Then the relation

$$\langle \mathbf{x}, \mathbf{b} \rangle = \langle \mathbf{y}, \mathbf{a} \rangle \qquad (2.3.5)$$

---

† Our use of the term "adjoint" here is not really precise. Nevertheless, this usage is fairly standard and we shall continue to say that equation pairs such as (2.3.1) and (2.3.4) are adjoint to each other.

follows easily by taking the inner product of Eq. (2.3.1) with $\mathbf{y}$ and comparing with the inner product of Eq. (2.3.4) and $\mathbf{x}$. We shall use these facts in the discussion to follow.

We assume now that the matrix $|H|$ of absolute values of H has spectral radius† less than unity. In the application to the multigroup transport equation we shall attempt to justify this assumption on physical grounds. For the moment, however, it must be regarded as a restriction on the type of algebraic problem to be solved. Since $|H|$ has spectral radius less than unity, so does $|H'|$ because $|H|$ and $|H'|$ have the same eigenvalues.

First consider an arbitrary discrete random walk process on the states $1, \ldots, N$:

$$p_i^1 = P(i_1 = i),$$

$$p_{ij} = P(i_{m+1} = i \mid i_m = j \text{ and } k > m), \qquad (2.3.6)$$

$$p_j = P(k = m \mid i_m = j) = 1 - \sum_{i=1}^{N} p_{i,j},$$

subject to the conditions $h_{ij} \neq 0 \Rightarrow p_{ij} \neq 0$ and $a_i \neq 0 \Rightarrow p_i^1 \neq 0$. We also assume that the matrix $K = (p_{ij})$ has spectral radius less than unity. It follows fairly easily from this latter assumption that the number of collisions made until termination is finite with probability one. As a consequence of this, if ξ is a random variable on the space of all discrete random walks α, its expected value, $E[\xi]$, may be written

$$E[\xi] = \sum_{\alpha} P(\alpha)\xi(\alpha) = \sum_{k=1}^{\infty} \sum_{\alpha \in \Lambda_k} P(\alpha)\xi(\alpha),$$

where Λ_k is the set of random walks terminating in exactly k collisions. In other words, the assumption $\rho(K) < 1$ is necessary to guarantee that particles which collide an infinite number of times may be ignored for the purpose of calculating expected values of random variables of interest. In Chapter 3 we shall discuss this point much more fully, and for continuous, rather than discrete, processes. Here, however, we are concerned mainly with motivating the discussion in Chapter 3 and we content ourselves with a more formal derivation, avoiding detailed probability arguments.

Let $\alpha = (i_1, \ldots, i_k)$ be an arbitrary random walk beginning at i_1 and terminating at i_k, and let

$$w_{ij} = \begin{cases} h_{ij}/p_{ij} & \text{if } p_{ij} \neq 0, \\ 0 & \text{if } p_{ij} = 0. \end{cases} \qquad (2.3.7)$$

---

† The spectral radius of a matrix A, denoted $\rho(A)$, is the largest modulus of all eigenvalues of A. Thus it is the radius of the smallest circle in the complex plane which contains all eigenvalues of A.

Define a random variable on the space of walks α by

$$W_k(\alpha) = \frac{a_{i_1}}{p_{i_1}^\alpha} w_{i_2 i_1} \cdots w_{i_k i_{k-1}}. \tag{2.3.8}$$

Theorem 2.1 The random variable $\xi_{j_0}(\alpha) = W_k(\alpha)\delta_{i_k j_0}/p_{i_k}$ is an unbiased estimator of x_{j_0}, the (j_0)-th component of the solution of Eq. (2.3.1).

Proof. The statement of the theorem means, of course, that the random variable ξ_{j_0} has, as its expected value, the number x_{j_0}. This is so because

$$E[\xi_{j_0}(\alpha)] = \sum_\alpha P(\alpha)\xi_{j_0}(\alpha)$$

$$= \sum_{k=1}^\infty \sum_{i_k} \cdots \sum_{i_1} \frac{a_{i_1}}{p_{i_1}^1} w_{i_2 i_1} \cdots w_{i_k i_{k-1}} p_{i_1}^1 p_{i_2 i_1} \cdots p_{i_k i_{k-1}} p_{i_k} \delta_{i_k j_0}/p_{i_k}$$

$$= \sum_{k=1}^\infty \sum_{i_k} \cdots \sum_{i_1} a_{i_1} h_{i_2 i_1} \cdots h_{i_k i_{k-1}} \delta_{i_k j_0}$$

$$= a_{j_0} + [Ha]_{j_0} + [H^2 a]_{j_0} + \cdots$$

$$= x_{j_0},$$

provided the Neumann series converges absolutely. This convergence is assured by the assumption on the spectral radius of $|H|$, $\rho(|H|) < 1$.

If we apply this method to the system of Eqs. (2.2.14) with

$$p_i^1 = \frac{Q_i}{\sum_{i=1}^G Q_i} = a_i, \qquad p_{ij} = \frac{\Sigma_s^{ij}}{\Sigma_t^j} = h_{ij}, \qquad p_j = 1 - \sum_{i=1}^G p_{ij} = \frac{\Sigma_a^j}{\Sigma_t^j},$$

then $w_{ij} = 1$ if $p_{ij} \neq 0$ and $w_{ij} = 0$ if $p_{ij} = 0$. Also, $W_k(\alpha) = 1$ for every k and every α and

$$\xi_{j_0}(\alpha) = \delta_{i_k j_0}/p_{i_k}$$

$$= \delta_{i_k j_0} \frac{\Sigma_t^{i_k}}{\Sigma_a^{i_k}}$$

$$= \begin{cases} \dfrac{\Sigma_t^{j_0}}{\Sigma_a^{j_0}} & \text{if } i_k = j_0, \\ 0 & \text{if } i_k \neq j_0. \end{cases}$$

We have $E[\xi_{j_0}(\alpha)] = P_{j_0}$, the (j_0)-th component of the discrete collision density. The discrete absorption rate in state j_0 is defined by

$$A_{j_0} = \frac{\Sigma_a^{j_0}}{\Sigma_t^{j_0}} P_{j_0}, \tag{2.3.9}$$

since $\Sigma_a^{j_0}/\Sigma_t^{j_0}$ is the probability of absorption upon collision in state j_0 and P_{j_0}

is the expected number of collisions. We see that the simple binomial random variable

$$W_k(\alpha)\delta_{i_k j_0} = \begin{cases} 1 & \text{if } i_k = j_0, \\ 0 & \text{if } i_k \neq j_0 \end{cases} \qquad (2.3.10)$$

is an unbiased estimator of A_{j_0}. The random variable (2.3.10) simply counts the number of particles which are absorbed in state j_0 and is in some sense the simplest possible estimator of A_{j_0}.

More generally, if Σ^{j_0} is the cross section for any process in state j_0, the random variable

$$W_k(\alpha)\delta_{i_k j_0}\frac{\Sigma^{j_0}}{\Sigma_a^{j_0}} = \begin{cases} \dfrac{\Sigma^{j_0}}{\Sigma_a^{j_0}} & \text{if } i_k = j_0, \\ 0 & \text{if } i_k \neq j_0 \end{cases} \qquad (2.3.11)$$

is an unbiased estimator of the reaction rate

$$\frac{\Sigma^{j_0}}{\Sigma_t^{j_0}} P_{j_0}.$$

The process we have just described for estimating the components of the solution of a matrix equation will presently be shown to be the "adjoint" of one proposed by von Neumann and Ulam and first described by Forsythe and Leibler (Ref. 1). A modification of the von Neumann-Ulam estimator, suggested by Wasow (Ref. 2), involves the random variable

$$\eta_{j_0}(\alpha) = \sum_{m=1}^{k} W_m(\alpha)\delta_{i_m j_0}, \qquad (2.3.12)$$

which makes use of every collision in the state i_k, not just the last collision.

Theorem 2.2 The random variable η_j is an unbiased estimator of x_{j_0}, the (j_0)-th component of the solution of Eq. (2.3.1).

Proof.

$$E[\eta_{j_0}(\alpha)] = \sum_{k=1}^{\infty} \sum_{i_k} \cdots \sum_{i_1} \left(\sum_{m=1}^{k} W_m(\alpha)\delta_{i_m j_0} \right) p_{i_1}^1 p_{i_2 i_1} \cdots p_{i_k i_{k-1}} p_{i_k}$$

$$= \sum_{m=1}^{\infty} \sum_{k=m}^{\infty} \sum_{i_k} \cdots \sum_{i_1} W_m(\alpha)\delta_{i_m j_0} p_{i_1}^1 p_{i_2 i_1} \cdots p_{i_k i_{k-1}} p_{i_k}$$

$$= \sum_{m=1}^{\infty} \sum_{i_m} \cdots \sum_{i_1} W_m(\alpha)\delta_{i_m j_0} p_{i_1}^1 p_{i_2 i_1} \cdots p_{i_m i_{m-1}}$$

$$\left\{ p_{i_m} + \sum_{i_{m+1}} p_{i_{m+1} i_m} p_{i_{m+1}} + \sum_{i_{m+2}} \sum_{i_{m+1}} p_{i_{m+2} i_{m+1}} p_{i_{m+1} i_m} p_{i_{m+2}} + \cdots \right\}.$$

Now we may make use of the relations

$$p_{i_{m+1}} = 1 - \sum_{i_{m+2}} p_{i_{m+2} i_{m+1}} \qquad \text{(Eq. (2.2.2))}$$

to collapse the term in brackets to $(1 - \lim_{t \to \infty} P^t)$, where P^t is the t-th power of the transition matrix. But $\lim_{t \to \infty} P^t = 0$, since $P = H$ was assumed to have spectral radius less than unity.† Replacing the bracket by 1 we find that the series representation of $E[\eta_{j_0}(\alpha)]$ is the same as that of $E[\xi_{j_0}(\alpha)]$, which proves that η_{j_0} is an unbiased estimator of x_{j_0}.

We now attempt to justify the assumption that $\rho(P) < 1$. According to Ref. 3, P will be convergent (i.e. will have spectral radius less than one) if it is irreducible and if some column sum is strictly less than one, i.e. if $\sum_{i=1}^{G} p_{ij} < 1$ for some j. Physically, this means that it is ultimately possible to pass from any state to any other state (irreducibility) and that the probability of absorption is not zero in at least one state. This means that there is at least one eventual sink for the neutrons, and in the discrete case, at least, this is sufficient to guarantee a convergent transition matrix and, therefore, one whose Neumann series is convergent. The formulation of physically reasonable assumptions which would guarantee a convergent Neumann series in the continuous case is more difficult and will be discussed at some length in Chapter 3.

So far we have developed two methods for estimating a component of the solution of a matrix equation. We shall now develop two others, patterned after the estimators already discussed, but based on the adjoint matrix equation. Consider the equation

$$\mathbf{y} = H'\mathbf{y} + \boldsymbol{\delta}_{j_0} \qquad (2.3.13)$$

adjoint to Eq. (2.3.1), where $\boldsymbol{\delta}_{j_0}$ is the vector whose i-th component is the Kronecker delta, δ_{ij_0}. We notice that the solution vector $\mathbf{y}$ may be identified with the (j_0)-th column of the matrix

$$[I - H']^{-1} = I + H' + (H')^2 + \cdots,$$

the series being absolutely convergent since $\rho(|H'|) = \rho(|H|) < 1$.

Now choose an arbitrary discrete random walk process Eq. (2.3.6) subject to the conditions $\rho(K) < 1$, $h'_{ij} = h_{ji} \neq 0 \Rightarrow p_{ij} \neq 0$ and let $p^1_{j_0} \neq 0$ and

$$w_{ij} = \begin{cases} h_{ji}/p_{ij} & \text{if } p_{ij} \neq 0, \\ 0 & \text{if } p_{ij} = 0. \end{cases} \qquad (2.3.14)$$

The random variable (2.3.8) becomes

$$W_k(\alpha) = \frac{\delta_{i_1 j_0}}{p^1_{i_1}} w_{i_2 i_1} \cdots w_{i_k i_{k-1}}. \qquad (2.3.15)$$

Then we have

Theorem 2.3 The random variable

$$\xi_{l_0}(\alpha) = W_k(\alpha) \delta_{i_k l_0}/p_{i_k}$$

is an unbiased estimator of y_{l_0}, the (l_0)-th component of $\mathbf{y}$.

---

† See Ref. 3 for this result.

The proof of Theorem 2.3 is the same as that of Theorem 2.1. Now consider the random variable

$$\theta(\alpha) = \sum_{l_0=1}^{N} \xi_{l_0}(\alpha)a_{l_0} = W_k(\alpha)a_{i_k}/p_{i_k}, \tag{2.3.16}$$

where the a_{l_0} are defined by Eq. (2.3.1). We have

Theorem 2.4 The random variable $\theta(\alpha)$ is an unbiased estimator of the inner product

$$\langle \mathbf{y}, \mathbf{a} \rangle = \sum_{i=1}^{N} y_i a_i.$$

Proof.

$$E[\theta(\alpha)] = \sum_{l_0=1}^{N} a_{l_0} E[\xi_{l_0}(\alpha)] = \sum_{l_0=1}^{N} a_{l_0} y_{l_0}, \text{ proving the theorem.}$$

Now, by Eq. (2.3.5),

$$\langle \mathbf{y}, \mathbf{a} \rangle = \langle \mathbf{x}, \boldsymbol{\delta}_{j_0} \rangle = x_{j_0},$$

where $\mathbf{x}$ is the solution of Eq. (2.3.1). Therefore, the random variable (2.3.16) also is an unbiased estimator of x_{j_0}. It is this random variable which von Neumann and Ulam proposed (Ref. 1), and the adjoint of Eq. (2.3.12), namely,

$$\phi_{j_0}(\alpha) = \sum_{m=1}^{k} W_m(\alpha)a_{i_m}, \tag{2.3.17}$$

which was suggested by Wasow.

We observe that both estimators (2.3.16) and (2.3.17) score only for random walks which originate in the state j_0. As in the normal process for estimating x_{j_0} described earlier, the variable (2.3.16) scores only on the last collision, while (2.3.17) scores on every collision.

If we apply these considerations to the system of Eqs. (2.2.14), the adjoint equation may be written

$$\hat{\mathbf{P}} = \boldsymbol{\delta}_{j_0} + K'\hat{\mathbf{P}}. \tag{2.3.18}$$

The physical interpretation of Eq. (2.3.18) is no longer as convenient as for Eq. (2.2.14). However, a discrete random walk process, patterned after that of Eq. (2.2.14), may still be based on Eq. (2.3.18). We notice that Eq. (2.3.18) suggests that particles should always originate in state j_0 and move from state j to state i with probabilities p_{ij} proportional to $(K')_{ij}$. If we further impose the natural restraint

$$\sum_{i=1}^{N} p_{ij} = \frac{\Sigma_s^j}{\Sigma_t^j},$$

then we are led uniquely to the choices

$$
\begin{cases}
p_i^1 = \delta_{i j_0}, \\[2ex]
p_{ij} = \dfrac{\Sigma_s^{ji}}{\Sigma_s^{ji}} \dfrac{\Sigma_s^j}{\Sigma_t^{ji}} \dfrac{\Sigma_t^i}{\Sigma_t^i} = (K')_{ij} \dfrac{\Sigma_t^i}{\sum_{i=1}^{N} \Sigma_s^{ji}} \dfrac{\Sigma_s^j}{\Sigma_t^j}, \\[2ex]
p_j = 1 - \sum_{i=1}^{N} p_{ij} = 1 - \dfrac{\Sigma_s^j}{\Sigma_t^j} = \dfrac{\Sigma_a^j}{\Sigma_t^j}.
\end{cases}
\tag{2.3.19}
$$

With this random walk process, the weight Equation (2.3.14) becomes

$$
w_{ij} = \begin{cases}
\left(\dfrac{\Sigma_t^i}{\sum_{i=1}^{N} \Sigma_s^{ji}} \dfrac{\Sigma_s^j}{\Sigma_s^j} \right)^{-1} & \text{if} \quad p_{ij} \neq 0, \\[2ex]
0 & \text{if} \quad p_{ij} = 0
\end{cases}
\tag{2.3.20}
$$

and the random variable $\theta(\alpha)$ of Eq. (2.3.16) becomes

$$
\theta(\alpha) = \left[\frac{\Sigma_t^{i_1}}{\sum_{i=1}^{N} \Sigma_s^{j_0 i}} \frac{\Sigma_s^{j_0}}{\Sigma_t^{j_0}} \cdot \frac{\Sigma_t^{i_2}}{\sum_{i=1}^{N} \Sigma_s^{i_1 i}} \frac{\Sigma_s^{i_1}}{\Sigma_t^{i_1}} \cdots \frac{\Sigma_t^{i_k}}{\sum_{i=1}^{N} \Sigma_s^{i_k i}} \frac{\Sigma_s^{i_{k-1}}}{\Sigma_t^{i_{k-1}}} \right]^{-1} a_{i_k} \frac{\Sigma_t^{i_k}}{\Sigma_a^{i_k}}, \tag{2.3.21}
$$

which may be regarded as a product of weighting factors acquired upon collision and upon termination in the adjoint process. It appears that, in practice, estimators such as Eq. (2.3.21) fluctuate a good deal from random walk to random walk and therefore have rather large variances. We shall later discuss, for the integral equation case, transformations of the adjoint equations which effectively reduce the weighting factors in Eq. (2.3.21) to unity and result in the elimination of some of the variance in the estimator (2.3.21).

In the discussion above we have perhaps given the impression that non-unit weights are entirely undesirable in any Monte Carlo calculation. We shall later (see Section 3.7) discuss importance sampling in which nonunit weights are essential to the design of a low-variance scheme. The undesirability of the random variable (2.3.21) is not so much a result of the non-unit weighting factors as it is of the fact that these weights in no way seem to reflect the importance of the event. We hope that this distinction will be clarified when we discuss importance sampling in Section 3.7.

The random variable $\phi_{j_0}(\alpha)$ of Eq. (2.3.17) may also be applied to yield an unbiased estimator of each component y_{l_0} of the adjoint solution, and, hence, by linearity and Eq. (2.3.5), of x_{j_0}. It has the same advantages when used to solve adjoint equations as when used to solve the original equations.

We have now described four random variables suitable for estimating a given component x_{j_0} of the solution $\mathbf{x}$ of Eq. (2.3.1), two based on Eq. (2.3.1) itself, and two based on the adjoint Eq. (2.3.13). We now examine these estimators in somewhat greater detail.

We observe that when the discrete random walk process is based on the normal Eq. (2.3.1), a source may be constructed from the known vector **a** and the matrix H may be used to control weights upon transition from state to state in the transport process which results. When the random variable ξ_{j_0} is used in conjunction with such a process, j_0 is the index at which termination takes place, so that each particle history contributes information to the estimation of *some* component of **x**. A similar reasoning is valid when the estimator η_{j_0} is used; here each particle history contributes information to some component of **x** on every collision it makes. One might expect that the estimator η_{j_0} always has a smaller variance than ξ_{j_0}, but this is not so. For example, consider the system Eq. (2.2.14) with Q_i arbitrary, $p_{ij} = \Sigma_s^{ij}/\Sigma_t^{j} = h_{ij}$, $p_j = 0$ for all j except $j = j_0$. Then $\xi_{j_0}(\alpha) = \Sigma_t^{j_0}/\Sigma_a^{j_0}$ for every α, since every walk must terminate in state j_0. Since the random variable ξ_{j_0} has the same value for every history, it gives a zero variance estimate of the discrete collision density in state j_0. Notice that the condition $p_j = 0$ for $j \neq j_0$ implies that $\Sigma_s^{j} = \Sigma_t^{j}$ for $j \neq j_0$ and this implies that $x_{j_0} = \Sigma_t^{j_0}/\Sigma_a^{j_0}$. The estimator η_{j_0}, on the other hand, reduces to

$$\eta_{j_0}(\alpha) = k,$$

where k is the number of collisions made in state j_0 by the history α. Clearly, η_{j_0} has a nonzero variance, so, for this very special problem, the variance of ξ_{j_0} is smaller than that of η_{j_0}. Wasow (Ref. 2) has described a condition under which η_{j_0} will have smaller variance than ξ_{j_0}. The result is the following.

Theorem 2.5 Suppose that $a_{j_0} = 1$ and $a_i = 0$ for $i \neq j_0$. Then var $[\eta_{j_0}] \leq$ var $[\xi_{j_0}]$ if and only if

$$p_{j_0} \leq \frac{\nu_{j_0}}{2 - \nu_{j_0}},$$

where ν_{j_0} is the probability that a discrete random walk starting at j_0 never returns to the state j_0.

Notice that in the special problem cited above, $\nu_{j_0} = 0$ and $p_{j_0} > 0$.

When the discrete random walk process is based on the adjoint Eq. (2.3.18), as in Eq. (2.3.19), then every particle must originate at the index j_0. Thus, in the adjoint process, all of the information is used to estimate the single component x_{j_0}. It seems natural to expect that this is a more efficient way of obtaining estimates of only a single component of the solution, and this is generally so. However, when knowledge of every component of **x** is desired, then it is necessary to base the random walk process on Eq. (2.3.1). This multiplicity of available solutions is a common occurrence in Monte Carlo; the greatest efficiency is obtained when the method used is designed for the specific class of problems to be solved. Often this must be done at the cost of a certain amount of generality.

All of the methods previously discussed may easily be extended to the problem of estimating the inner product of the solution vector $\mathbf{x}$ with a known vector. Since this problem is a common one in the integral equation case, we shall indicate how this extension can be accomplished.

Suppose it is desired to estimate the inner product

$$\langle \mathbf{x}, \mathbf{g} \rangle, \tag{2.3.22}$$

where $\mathbf{g}$ is a known vector and $\mathbf{x}$ satisfies

$$\mathbf{x} = H\mathbf{x} + \mathbf{a}. \tag{2.3.23}$$

The equation adjoint to the pair (2.3.22), (2.3.23) is

$$\mathbf{y} = H'\mathbf{y} + \mathbf{g} \tag{2.3.24}$$

and, as before, the relationship

$$\langle \mathbf{x}, \mathbf{g} \rangle = \langle \mathbf{y}, a \rangle \tag{2.3.25}$$

is easily established. Now, the duality between Eqs. (2.3.23) and (2.3.24) is more apparent than in the special case $\mathbf{g} = \boldsymbol{\delta}_{j_0}$ treated already. If the discrete random walk process is based on Eq. (2.3.23), the components x_i of $\mathbf{x}$ may be estimated by either the random variables ξ_i or η_i, and the random variable $\sum_{i=1}^{N} \xi_i g_i$, or $\sum_{i=1}^{N} \eta_i g_i$, provides an unbiased estimate of $\langle \mathbf{x}, \mathbf{g} \rangle$. On the other hand, if the random walk process is based on the adjoint Eq. (2.3.24), the components y_i may be similarly estimated and the random variables $\sum_{i=1}^{N} \xi_i a_i$ and $\sum_{i=1}^{N} \eta_i a_i$ give unbiased estimates of the same inner product. Notice that the special case $\mathbf{g} = \boldsymbol{\delta}_{j_0}$ reduces to the examples previously treated. The relative efficiencies of the direct and adjoint methods depend on the specific nature of the vector $\mathbf{g}$.

2.4 SOLUTIONS OF INTEGRAL EQUATIONS

In this section and the next we shall discuss methods for estimating various characteristics of the solution of Fredholm integral equations of the second kind. Such equations arise in the treatment of neutron transport and so provide the primary application at which this book is aimed. As we shall see in Section 2.5, some of the methods discussed in the previous section for matrix equations generalize directly to integral equations while others become meaningless for integral equations.

In order to maintain as close a connection as possible with the discrete case previously treated, we first postulate a phase space Γ which is a six-dimensional continuum, three independent coordinates being used to represent position, and three to indicate speed and direction. If the symbols x, x_k, etc.

are used to denote generic points of such a phase space, then a particle random walk is characterized by first collision† in some state x_1, subsequent transmission through a sequence of states $x_2, x_3, \ldots$, and, finally, termination with probability one in some state x_k. A continuous nonmultiplying random walk process is therefore completely specified by functions

$$p^1(x) = P(x_1 = x),$$
$$p(x, y) = P(x_{n+1} = x \mid x_n = y \text{ and } k > n), \qquad (2.4.1)$$
$$p(x) = P(k = n \mid x_n = x)$$

with the requirements that

$$p^1(x) \geq 0, \qquad p(x, y) \geq 0 \qquad \text{and}$$

$$\int_\Gamma p^1(x) \, dx = 1, \qquad (2.4.2)$$

$$\int_\Gamma p(x, y) \, dx = 1 - p(y) \leq 1 \qquad \text{for all } y \in \Gamma.$$

Let $P^n(x)$ denote the probability density of making collision n at point $x, n = 1, \ldots$. Then

$$P^n(x) = P(x_n = x \mid k > n - 1, n \geq 2),$$
$$P^1(x) = P(x_1 = x) = p^1(x). \qquad (2.4.3)$$

Clearly

$$P^n(x) = \int_\Gamma p(x, y) P^{n-1}(y) \, dy, \qquad n \geq 2. \qquad (2.4.4)$$

By analogy with the discrete case, we may define a random variable $X(x)$ by $X(x) =$ number density of collisions made at x. Then X is a discrete random variable on the space of (continuous) random walks whose expected value $E[X]$ may be calculated by

$$E[X] = 1 \cdot P^1(x) + 1 \cdot P^2(x) + \cdots$$

$$= \sum_{n=1}^{\infty} P^n(x). \qquad (2.4.5)$$

---

† As in the discrete case (see discussion on p. 44), we need not yet concern ourselves with the birth process or the process which transfers particles from the birth point to a first collision point. Only the density of first collisions enters our equations here.

We call $P(x) = E[X(x)]$ the *continuous collision density*, or simply *collision density*, at x, and, making use of Eq. (2.2.4),

$$\sum_{n=2}^{\infty} P^n(x) = \sum_{n=2}^{\infty} \int_\Gamma p(x, y) P^{n-1}(y) \, dy$$

$$= \int_\Gamma p(x, y) \sum_{n=2}^{\infty} P^{n-1}(y) \, dy$$

$$= \int_\Gamma p(x, y) \sum_{n=1}^{\infty} P^n(y) \, dy.$$

Therefore we have arrived at the integral equation for $P(x)$,

$$P(x) = p^1(x) + \int_\Gamma p(x, y) P(y) \, dy. \tag{2.4.6}$$

Notice the similarity between this equation and the corresponding matrix equation (2.2.8) for a discrete process. Notice also that in Eq. (2.4.6) the kernel $p(x, y)$ is used to describe transitions from y to x, rather than from x to y. This is consistent with conventions adopted earlier for matrix equations (see second footnote, p. 42).

The physical process that we shall treat in the remainder of this book, namely, the motion of neutrons in a reactor, is described by an integral equation of the type (2.4.6). In fact, the integral equation for the neutron collision density is (see Ref. 4)

$$\psi(\mathbf{r}, \mathbf{E}) = \iint \psi(\mathbf{r}', \mathbf{E}') K(\mathbf{r}, \mathbf{E}; \mathbf{r}', \mathbf{E}') \, d\mathbf{E}' \, d\mathbf{r}' + \int Q(\mathbf{r}', \mathbf{E}) T(\mathbf{r}', \mathbf{r}; \mathbf{E}) \, d\mathbf{r}', \tag{2.4.7}$$

where $\mathbf{x} = (\mathbf{r}, \mathbf{E})$ separates the spatial vector $\mathbf{r}$ from the velocity $\mathbf{E}$. The kernel K factors into the form

$$K(\mathbf{r}, \mathbf{E}; \mathbf{r}', \mathbf{E}') = C(\mathbf{E}', \mathbf{E}; \mathbf{r}') T(\mathbf{r}', \mathbf{r}; E), \tag{2.4.8}$$

where C is the collision kernel, T is the transport kernel and Q is the physical neutron source density. The kernel K gives the density of particle sentering collision at $(\mathbf{r}', \mathbf{E}')$ which enter their next collision at $(\mathbf{r}, \mathbf{E})$. The kernel T may be characterized by stating that, for a particle leaving a collision (or the source) at $(\mathbf{r}', \mathbf{E})$, the expected number of next collisions in the spatial volume V is

$$\int_V T(\mathbf{r}', \mathbf{r}; \mathbf{E}) \, d\mathbf{r}.$$

The collision kernel C is similarly defined so that, for a particle entering a collision at $(\mathbf{r}, \mathbf{E}')$ the expected number of particles leaving the collision in the energy volume V is

$$\int_V C(\mathbf{E}', \mathbf{E}; \mathbf{r}) \, d\mathbf{E}.$$

Explicitly, if $\boldsymbol{\omega}$ is a unit vector in the direction of $\mathbf{E}$, $T(\mathbf{r}', \mathbf{r}; \mathbf{E})$ may be written

$$T(\mathbf{r}', \mathbf{r}; \mathbf{E}) = \Sigma_t(\mathbf{r}, \mathbf{E}) \exp \left[- \int_0^{\boldsymbol{\omega} \cdot (\mathbf{r} - \mathbf{r}')} \Sigma_t(\mathbf{r}' + s\boldsymbol{\omega}, \mathbf{E}) \, ds \right] \qquad (2.4.9)$$

for all $\mathbf{r}$ such that $\mathbf{r} - \mathbf{r}'$ is parallel to $\boldsymbol{\omega}$ and $\boldsymbol{\omega} \cdot (\mathbf{r} - \mathbf{r}') \geq 0$. For example, if $\Sigma_t(\mathbf{r}', E)$, the total macroscopic cross section at $(\mathbf{r}', E)$, is spatially constant along the direction $\boldsymbol{\omega}$ (as in an infinite medium), Eq. (2.4.9) reduces to

$$T(\mathbf{r}', \mathbf{r}; \mathbf{E}) = \Sigma_t(E) \exp \left(-\Sigma_t(E)d \right), \qquad (2.4.10)$$

where d is the distance from $\mathbf{r}'$ to $\mathbf{r}$.

The collision kernel $C(\mathbf{E}', \mathbf{E}; \mathbf{r})$ may be represented explicitly as

$$C(\mathbf{E}', \mathbf{E}; \mathbf{r}) = \sum_i p_i C_i(\mathbf{E}', \mathbf{E}; \mathbf{r}), \qquad (2.4.11)$$

where p_i denotes the probability of a scattering collision of type i and C_i is the corresponding collision kernel. Each C_i is normalized to the mean number of secondaries, ν, per event:

$$\int C_i(\mathbf{E}', \mathbf{E}; \mathbf{r}) \, d\mathbf{E} = \nu_i. \qquad (2.4.12)$$

For elastic and most† inelastic scattering events, $\nu = 1$, while for fission, $\nu > 1$. The numbers p_i may be written as

$$p_i = \frac{\Sigma_{s_i}(\mathbf{r}, \mathbf{E}')}{\Sigma_t(\mathbf{r}, \mathbf{E}')}, \qquad (2.4.13)$$

where Σ_{s_i} is the macroscopic scattering cross section for a scattering of type i. If there is no fission, i.e. in a non-multiplying medium, then

$$\sum_i p_i = \frac{\Sigma_s(\mathbf{r}, \mathbf{E}')}{\Sigma_t(\mathbf{r}, \mathbf{E}')}, \qquad (2.4.14)$$

where $\Sigma_s(\mathbf{r}, \mathbf{E}') = \sum_i \Sigma_{s_i}(\mathbf{r}, \mathbf{E}')$ is the macroscopic scattering cross section for all types of scattering events. The form of each C_i is determined by the physics of the problem being solved as well as mathematical approximations which are frequently desirable and useful. For example, it is often useful to assume that

$$C_i(\mathbf{E}', \mathbf{E}; \mathbf{r}) = \frac{1}{4\pi} f(E', E; \mathbf{r}) \qquad (2.4.15)$$

is isotropic for all energies E'. Then

$$\int f(E', E; \mathbf{r}) \, dE = 1 \qquad (2.4.16)$$

---

† Here we ignore neutron multiplication by means of, for example, inelastic $(n, 2n)$-reactions.

and the function f is used to describe the distribution of speeds following collision. It follows from the above considerations that†

$$\iint K(\mathbf{r}, \mathbf{E}; \mathbf{r}', \mathbf{E}') \, d\mathbf{r} \, d\mathbf{E} = c(\mathbf{r}', \mathbf{E}'),$$

the mean number of secondaries per primary upon collision at $(\mathbf{r}', \mathbf{E}')$.

A frequently used integro-differential equation for the neutron flux $F(\mathbf{r}, \mathbf{E})$ is

$$\nabla \cdot \boldsymbol{\omega} F(\mathbf{r}, \mathbf{E}) + \Sigma_t(\mathbf{r}, E)F(\mathbf{r}, \mathbf{E}) = \int \Sigma_t(\mathbf{r}, E')F(\mathbf{r}, \mathbf{E}')C(\mathbf{E}', \mathbf{E}; \mathbf{r}) \, d\mathbf{E}' + Q(\mathbf{r}, \mathbf{E}),$$

$$(2.4.17)$$

where $\boldsymbol{\omega}$ is a unit vector in the direction of $\mathbf{E}$, $\Sigma_t(\mathbf{r}, E)$ is the total macroscopic cross section, and the collision density and flux are related via

$$\psi(\mathbf{r}, \mathbf{E}) = \Sigma_t(\mathbf{r}, E)F(\mathbf{r}, \mathbf{E}). \qquad (2.4.18)$$

Equations such as (2.4.7) and (2.4.17) form the mathematical model for the applications we shall discuss in the remainder of this monograph.

So far we have exhibited an integral equation (2.4.7) and an integro-differential equation (2.4.17) which describe the transport of neutrons. We see that the integro-differential equation (2.4.17) has the physical source density, Q, as its source term, while the integral equation (2.4.7) has as its source term the density of first collisions. It is often convenient to work with an integral equation whose source is the physical density of births, Q, rather than the density of first collisions. Such as integral equation may be obtained as follows.

We write the collision density $\psi(\mathbf{r}, \mathbf{E})$ as

$$\psi(\mathbf{r}, \mathbf{E}) = \int \chi(\mathbf{r}', \mathbf{E})T(\mathbf{r}', \mathbf{r}; \mathbf{E}) \, d\mathbf{r}', \qquad (2.4.19)$$

so that, from (2.4.7),

$$\chi(\mathbf{r}, \mathbf{E}) = \int \psi(\mathbf{r}, \mathbf{E}')C(\mathbf{E}', \mathbf{E}; \mathbf{r}) \, d\mathbf{E}' + Q(\mathbf{r}, \mathbf{E})$$

$$= \iint \chi(\mathbf{r}', \mathbf{E}')L(\mathbf{r}, \mathbf{E}; \mathbf{r}', \mathbf{E}') \, d\mathbf{r}' \, d\mathbf{E}' + Q(\mathbf{r}, \mathbf{E}), \qquad (2.4.20)$$

where

$$L(\mathbf{r}, \mathbf{E}; \mathbf{r}', \mathbf{E}') = T(\mathbf{r}', \mathbf{r}; \mathbf{E}')C(\mathbf{E}', \mathbf{E}; \mathbf{r}). \qquad (2.4.21)$$

We notice that the integral equation (2.4.20) is of the same type as Eq. (2.4.7),

---

† Strictly speaking, this normalization is valid only for an infinite medium or in a cell. However, we may continue to use this relationship, even in finite media, if we surround the region of interest with a black absorber and take the integral over all space. We shall adopt this convention here and throughout the remainder of the book.

except that its source term Q is the physical source density and its kernel has the collision and transport factors inverted. The kernel L gives the density of particles which leave a collision at $(\mathbf{r}, \mathbf{E})$, having left the previous collision (or source) at $(\mathbf{r}', \mathbf{E}')$. The solution, $\chi(\mathbf{r}, \mathbf{E})$, of Eq. (2.4.20) may be interpreted by stating that the expected number of particles appearing per unit time in a volume V of phase space, either directly from the source or from a collision, is $\int_V \chi(\mathbf{r},\mathbf{E})\, d\mathbf{r}\, d\mathbf{E}$. From the definition of T (Eq. 2.4.9) and the relation (2.4.19) it can be shown that

$$F(\mathbf{r}, \mathbf{E}) = \int_0^\infty \chi(\mathbf{r} - s\boldsymbol{\omega}, \mathbf{E}) \exp\left[-\int_0^s \Sigma_t(\mathbf{r} - t\boldsymbol{\omega}, \mathbf{E})\, dt \right] ds, \quad (2.4.22)$$

where F is the neutron flux. We also observe that the integro-differential Eq. (2.4.17) may be written

$$\nabla \cdot \boldsymbol{\omega} F(\mathbf{r}, \mathbf{E}) + \Sigma_t(\mathbf{r}, E)F(\mathbf{r}, \mathbf{E}) = \chi(\mathbf{r}, \mathbf{E}). \quad (2.4.23)$$

We notice that the definition (2.4.19), or the equivalent relationship Eq. (2.4.22), allows one to represent the dependent variable in the transport equation as a sum over a discrete index representing the collision number. That is, let

$$\psi(\mathbf{r}, \mathbf{E}) = \sum_{n=1}^\infty \psi_n(\mathbf{r}, \mathbf{E}) \quad (2.4.24)$$

and

$$\chi(r, E) = \sum_{n=0}^\infty \chi_n(\mathbf{r}, \mathbf{E}), \quad (2.4.25)$$

where ψ_n is the density of particles which have undergone $(n-1)$ collisions and χ_n is the density of particles appearing as a result of collision n ($n = 0$ designating birth). Then we have

$$\chi_0(\mathbf{r}, \mathbf{E}) = Q(\mathbf{r}, \mathbf{E}), \quad (2.4.26)$$

$$\chi_n(\mathbf{r}, \mathbf{E}) = \int \psi_n(\mathbf{r}, \mathbf{E}')C(\mathbf{E}', \mathbf{E}; \mathbf{r})\, d\mathbf{E}', \qquad n = 1, 2, \ldots, \quad (2.4.27)$$

$$\psi_{n+1}(\mathbf{r}, \mathbf{E}) = \int \chi_n(\mathbf{r}', \mathbf{E})T(\mathbf{r}', \mathbf{r}; \mathbf{E})\, d\mathbf{r}', \qquad n = 0, 1, 2\ldots. \quad (2.4.28)$$

Relationships such as these are useful to keep in mind and provide the tools needed to go back and forth between integral and integro-differential transport equations.

2.5 BASIC ESTIMATORS FOR INTEGRAL EQUATIONS

We shall presently observe a number of important differences between random variables useful in discrete random walk processes and those useful in the continuous case. For example, the estimators ξ_{j_0} and η_{j_0} applied to the normal matrix equation yield estimates of a single component x_{j_0} of the matrix equation. Their continuous analogs may not be used to obtain estimates of the collision density or flux at a single point; it is easy to see that such estimators tend to a Dirac delta function in the continuous case and thus become meaningless.† The basic difference between the discrete and continuous estimators is that in the discrete case a single component of the solution may be written as the inner product of the solution vector with the unit vector, all of whose components vanish except for the component in question. In the continuous case, however, the value of the solution function at a point may not be written as an inner product without resorting to the use of the delta function. We shall see that point estimators of the solution function may, nevertheless, be obtained by solving adjoint equations. When the adjoint equation is used, the delta function appears in the equation instead of in the estimator, where it would give rise to a statistically unreliable procedure.

We begin with estimators appropriate for inner products. In the continuous case the inner product in question is an integral

$$\langle \psi, g \rangle \equiv \iint_{\Gamma} \psi(\mathbf{r}, \mathbf{E}) g(\mathbf{r}, \mathbf{E}) \, d\mathbf{r} \, d\mathbf{E} \tag{2.5.1}$$

taken over all points of the phase space Γ. Suppose, then, that the unknown function ψ satisfies the integral equation

$$\psi(\mathbf{r}, \mathbf{E}) = \iint_{\Gamma} \psi(\mathbf{r}', \mathbf{E}') K(\mathbf{r}, \mathbf{E}; \mathbf{r}', \mathbf{E}') \, d\mathbf{r}' \, d\mathbf{E}' + S(\mathbf{r}, \mathbf{E}) \tag{2.5.2}$$

and that it is desired to estimate an integral of the form of Eq. (2.5.1). A rigorous development of the methods we shall discuss for estimating such quantities would depend on the definition of a probability space which embodies, in a mathematical sense, the properties of the random walk space we need. Such a construction is made in complete detail in Ref. 5 and some aspects of this construction are discussed in Chapter 3 at greater length. As we have discussed earlier in connection with the discrete case (see p. 48), in Chapter 2 we are interested more in motivation than in complete rigor. We therefore postpone difficulties connected with the abstract probability model until Chapter 3. Here we make the blanket assumption that, with probability one, all random walks terminate after a finite number of collisions. This permits expected values to be written as simple infinite series of integrals. Conditions which guarantee that this may be done will be established in Section 3.3.

---

† In the sense that they give an infinite contribution with probability zero.

Analogously to the discrete case, the integral equation adjoint to Eq. (2.5.2) and Eq. (2.5.1) is

$$\psi^*(\mathbf{r}, E) = \iint \psi^*(\mathbf{r}', E')K(\mathbf{r}', E'; \mathbf{r}, E)\, d\mathbf{r}'\, dE' + g(\mathbf{r}, E) \qquad (2.5.3)$$

and once again, the relationship

$$\langle \psi, g \rangle = \iint_{\Omega} \psi g\, d\mathbf{r}\, dE = \langle \psi^*, S \rangle = \iint_{\Omega} \psi^* S\, d\mathbf{r}\, dE \qquad (2.5.4)$$

holds, displaying the duality between the simulation of Eqs (2.5.2) and (2.5.3) in estimating inner products.

To evaluate $\langle \psi, g \rangle$, we first construct an arbitrary continuous random walk process $p^1(x)$, $p(x, y)$, $p(x)$ subject to the restrictions $K(x, y) \neq 0 \Rightarrow p(x, y) \neq 0$, $S(x) \neq 0 \Rightarrow p^1(x) \neq 0$ and $g(x) \neq 0 \Rightarrow p(x) \neq 0$. As discussed on p. 61, we also impose restrictions (delineated in Chapter 3) which guarantee that, with probability one, random walks terminate in a finite number of steps. Suppose $\alpha = (x_1, \ldots, x_k)$ is an arbitrary random walk beginning at x_1 and terminating at x_k. The random variable we consider is analogous to the one suggested by von Neumann in the discrete case, namely,

$$\xi(\alpha) = \frac{S(x_1)}{p^1(x_1)} w(x_2, x_1) \cdots w(x_k, x_{k-1}) \frac{g(x_k)}{p(x_k)}, \qquad (2.5.5)$$

where

$$w(x, y) = \begin{cases} K(x, y)/p(x, y) & \text{if } p(x, y) \neq 0, \\ 0 & \text{if } p(x, y) = 0. \end{cases} \qquad (2.5.6)$$

Theorem 2.6 If the random variable ξ is bounded,† its expected value is $\langle \psi, g \rangle$, i.e. ξ is an unbiased estimator of $\langle \psi, g \rangle$.

Proof. The boundedness of ξ guarantees that ξ has a finite expected value. Then we may write

$$E[\xi] = \sum_{\alpha} P(\alpha)\xi(\alpha)$$

$$= \sum_{k=0}^{\infty} \int \cdots \int dx_1 \cdots dx_k \frac{S(x_1)}{p^1(x_1)} w(x_2, x_1) \cdots w(x_k, x_{k-1})g(x_k)/p(x_k)$$

$$\cdot p^1(x_1)p(x_2, x_1) \cdots p(x_k, x_{k-1})p(x_k)$$

$$= \sum_{k=1}^{\infty} \int \cdots \int dx_1 \cdots dx_k S(x_1)K(x_2, x_1) \cdots K(x_k, x_{k-1})g(x_k)$$

$$= \int_{\Gamma} g(x)\psi(x)\, dx$$

---

† This restriction is relaxed slightly in Chapter 3.

provided the series of integrals converges, which we shall assume for the present. In Chapter 3 we shall establish conditions (see Theorem 3.5 and the discussion on p. 103) under which this convergence is assured. Thus, we see that the random variable ξ is an unbiased estimator of the reaction rate (2.5.1), proving the theorem.

If we define

$$W_m(\alpha) = \frac{S(x_1)}{p^1(x_1)} w(x_2, x_1) \cdots w(x_m, x_{m-1}), \qquad (2.5.7)$$

then the continuous analogue of the Wasow random variable is

$$\eta(\alpha) = \sum_{m=1}^{k} W_m(\alpha) g(x_m), \qquad (2.5.8)$$

which can also be shown to be an unbiased estimator of (2.5.1). (See discussion in Section 3.6, especially Theorem 3.5.)

If we apply these random variables to Eq. (2.5.2) with no fission and with the choices

$$p^1(x) = S(x), \qquad p(x, y) = K(x, y),$$
$$p(y) = 1 - \int_\Gamma K(x, y)\, dx = \frac{\Sigma_a(y)}{\Sigma_t(y)}, \qquad (2.5.9)$$

where Σ_a and Σ_t are the absorption and total cross sections, respectively. Then Eq. (2.5.5) becomes

$$\xi(\alpha) = g(x_k) \cdot \Sigma_t(x_k)/\Sigma_a(x_k), \qquad (2.5.10)$$

and Eq. (2.5.8) becomes

$$\eta(\alpha) = \sum_{m=1}^{k} g(x_m). \qquad (2.5.11)$$

If these random variables are used for estimating the absorption rate in some volume V of phase space, then

$$g(x) = \frac{\Sigma_a(x)}{\Sigma_t(x)} \chi_V(x), \qquad (2.5.12)$$

where

$$\chi_V(x) = \begin{cases} 1 & \text{if } x \in V, \\ 0 & \text{if } x \notin V \end{cases}$$

and Eq. (2.5.10) becomes

$$\xi(\alpha) = \begin{cases} 1 & \text{if } x_k \in V, \\ 0 & \text{if } x_k \notin V, \end{cases} \qquad (2.5.13)$$

while Eq. (2.5.11) becomes

$$\eta(\alpha) = \sum_{m=1}^{k} \frac{\Sigma_a(x_m)}{\Sigma_t(x_m)} \chi_{V}(x_m). \tag{2.5.14}$$

These basic estimators will be used repeatedly in later discussions.

We might also apply the preceding theory to the integral equation (2.4.20) with the choices†

$$p^1(x) = Q(x), \qquad p(x, y) = L(x, y), \qquad p(y) = 1 - \int L(x, y)\, dx. \tag{2.5.15}$$

Then the integral (2.5.1) becomes

$$\iint_{\Gamma} \psi g = \iiint \chi(\mathbf{r}', \mathbf{E}) T(\mathbf{r}', \mathbf{r}; \mathbf{E}) g(\mathbf{r}, \mathbf{E})\, d\mathbf{r}'\, d\mathbf{r}\, d\mathbf{E},$$

so that the random variables ξ and η of Eqs. (2.5.10) and (2.5.11) must be altered by the replacement of g with $\int T(\mathbf{r}', \mathbf{r}; \mathbf{E}) g(\mathbf{r}, \mathbf{E})\, d\mathbf{r}$. This point of view is often useful, as we shall see in Section 3.6.

The random walk process defined by Eq. (2.5.9) or Eq. (2.5.15) might properly be called the *analog random walk process* for a problem without multiplication. The execution of the analog random walk process provides a faithful simulation of the behavior of the neutrons in the steady state described by Eq. (2.5.2). The methods of Sections 1.5, 1.6, and 1.7 may be applied directly to Eq. (2.5.9) or Eq. (2.5.15) to construct sample random walks. The function $S(\mathbf{r}, \mathbf{E})$ is the density of first collisions and may sometimes be constructed analytically and then sampled directly. Otherwise, the physical source density $Q(\mathbf{r}, \mathbf{E})$ is sampled and the density S is constructed by transporting the particles from their point of origin to their first collision point by using the kernel T,

$$S(\mathbf{r}, \mathbf{E}) = \int Q(\mathbf{r}', \mathbf{E}) T(\mathbf{r}', \mathbf{r}; \mathbf{E})\, d\mathbf{r}. \tag{2.5.16}$$

Having arrived at the first collision point $(\mathbf{r}_1, \mathbf{E}_1)$, the kernel C is used to identify a scattering or absorbing event at $(\mathbf{r}_1, \mathbf{E}_1)$. A discrete distribution is used for this purpose, namely, the numbers p_i of Eq. (2.4.13). If the particle is not absorbed at $(\mathbf{r}_1, \mathbf{E}_1)$, an i is determined by using the p_i. Having fixed the type i of scattering event, the scattering kernel C_i is used to select a new direction and speed. Finally, the distance to the next event is determined by again sampling T. This process is repeated unless the particle is absorbed or else terminated because it may have left the region or energy range of interest.

Another frequently used random walk process is one which is identical to the analog process just described except for the fact that absorption is

---

† We defer a discussion of minor problems resulting from the use of the kernel L until Chapter 3 (see pp. 96ff).

forbidden. Such a process is defined by

$$p^1(x) = S(x),$$

$$p(x, y) = K(x, y)\Big/ \int_\Gamma K(x, y)\, dx = K(x, y)\Sigma_t(y)/\Sigma_s(y), \quad (2.5.17)$$

$$p(y) = 1 - \int p(x, y)\, dx = 0, \qquad \text{for all } y.$$

It is understood that when such a process is used, some artificial means of terminating histories—such as energy cutoff, or Russian roulette (see Section 3.8)—must be employed. The random variable (2.5.5) becomes meaningless since $p(y) = 0$ for all y. However, Eq. (2.5.8) may be used and becomes

$$\eta_1(\alpha) = \sum_{m=1}^{k}\left[\prod_{j=1}^{m}\left(\frac{\Sigma_s(x_j)}{\Sigma_t(x_j)}\right)\right]g(x_m). \qquad (2.5.18)$$

The quantity $\Sigma_s(x)/\Sigma_t(x)$ is the (analog) survival probability at x and the factor

$$\prod_{j=1}^{m}\frac{\Sigma_s(x_j)}{\Sigma_t(x_j)}$$

may be regarded as a multiplicative weight factor applied to the neutron history which is the probability of surviving collisions at $x_1, \ldots, x_m$. If the neutron is assigned a weight of unity at birth, then the random variable (2.5.18) may be interpreted as recording the quantity $g(x_m)$ upon collision at x_m for a neutron whose weight has been reduced by the factor

$$\prod_{j=1}^{m}\frac{\Sigma_s(x_j)}{\Sigma_t(x_j)}.$$

This notion of having a variable weight assigned to a neutron history is a useful one and is discussed more fully in the next section.

We observe that as the region V of phase space shrinks, both random variables ξ and η become less and less efficient, since fewer histories will terminate in V or suffer collisions in V. Furthermore, as observed earlier, the estimators ξ and η may not be used to estimate the solution ψ at a single point $\hat{x}$, since this would require that the function $g(x)$ tend to $\delta(x - \hat{x})$, which would render Eqs. (2.5.5) and (2.5.8) meaningless. However, the delta function may be used as a source density in the adjoint Eq. (2.5.3) (so that all particle histories are required to originate at $\hat{x}$) and the estimators ξ and η, applied to the adjoint process, become meaningful and useful.† As

---

† Estimators of a distinctly different character, such as the one suggested by Kalos (see Ref. 6 and Section 3.6), may also be used to obtain estimates of $\psi(\hat{x})$.

is suggested by this discussion, the adjoint estimators will tend to have smaller variances than the normal estimators as the volume V approaches a point.

The definition of a continuous random walk process based on the adjoint Eq. (2.5.3) is parallel to that defined by Eqs. (2.5.2), namely,

$$p^1(x) = g(x) \Big/ \int_\Gamma g(x)\, dx,$$

$$p(x, y) = \frac{K(y, x)}{\int_\Gamma K(y, x)\, dx} \frac{\Sigma_s(y)}{\Sigma_t(y)}, \qquad (2.5.19)$$

$$p(y) = 1 - \int_\Gamma p(x, y)\, dx = 1 - \frac{\Sigma_s(y)}{\Sigma_t(y)}.$$

With such a random walk process, the random variable $\xi(\alpha)$ of Eq. (2.5.5) becomes

$$\xi(\alpha) = \int g(x)\, dx \int_\Gamma K(x_1, x)\, dx \frac{\Sigma_t(x_1)}{\Sigma_s(x_1)} \cdots$$

$$\times \int_\Gamma K(x_{k-1}, x)\, dx \frac{\Sigma_t(x_{k-1})}{\Sigma_s(x_{k-1})} S(x_k) \frac{\Sigma_t(x_k)}{\Sigma_a(x_k)}, \quad (2.5.20)$$

which must again be interpreted as a product of weighting factors arising from the successive collisions of the history. Notice the similarity of Eq. (2.5.20) to Eq. (2.3.21).

2.6 THE TRACK LENGTH ESTIMATOR

In the preceding sections we have developed basic transport estimators which may be used to estimate activation rates in regions or, by using reciprocity, at a point. All of the estimators discussed so far may be thought of as recording *at* collision points of the neutron history. In nuclear design problems it is frequently the case that regions of extremely small dimension (measured in units of mean free path) occur, and in such regions neutron collisions are infrequent. For this reason, estimators which record only at collision points are likely to have large variances associated with them in such regions. In this section we shall develop estimators which may be regarded as recording continuously along the neutron flight path and which therefore have inherent advantages in optically thin regions. Such estimators do not really represent a substitute for the use of reciprocity in estimating activation rates at points or over regions of very small dimension. Indeed, as the region in question shrinks, the efficiency of track length estimators progressively degenerates, just as does the efficiency of collision estimators. Still, track length estimators provide good estimates of reaction rates over a surprisingly wide range of region dimension and are useful in conjunction with any Monte Carlo calculation, including ones based on reciprocity.

We shall introduce two related pairs of families of collision type estimators and apply a certain limiting process to each family. This limiting process may be regarded as one which gives rise to random variables which record information continuously along the flight path instead of only at discrete collision points. When applied to the first family, the limiting process yields the track length estimator. The same limiting process applied to members of the second family yields an exponential estimator suggested by Richtmyer (Ref. 7) for use in resonance escape calculations.

Assume, as usual, that it is desired to estimate a reaction rate R, represented by

$$R = \int_V g(x)\psi(x)\, dx, \tag{2.6.1}$$

where $\psi(x)$ is the collision density, satisfying Eq. (2.5.2), and V is any volume of phase space. By redefining $g(x)$ so that it vanishes outside of V, we may rewrite R as

$$R = \int_\Gamma g(x)\psi(x)\, dx, \tag{2.6.2}$$

the integral now extending over all of phase space Γ. In this form the function g is typically a ratio $\Sigma_r(x)/\Sigma_t(x)$ times the function $\chi_V(x)$,

$$\chi_V(x) = \begin{cases} 1, & x \in V, \\ 0, & x \notin V, \end{cases}$$

where $\Sigma_r(x)$ is the macroscopic cross section of some reaction. For example, if $\Sigma_r(x) = \Sigma_a(x)$, then R is the absorption rate in V. Equation (2.5.14) defines the random variable

$$\eta(\alpha) = \sum_{m=1}^{k} \frac{\Sigma_a(x_m)}{\Sigma_t(x_m)} \chi_V(x_m)$$

as an unbiased estimator of the absorption rate in V, with respect to the analog random walk process Eq. (2.5.9). More generally, the estimator

$$\eta_r(\alpha) = \sum_{m=1}^{k} \frac{\Sigma_r(x_m)}{\Sigma_t(x_m)} \chi_V(x_m) \tag{2.6.3}$$

is an unbiased estimator† of R with respect to the analog process. In Eq. (2.6.3) we continue to follow the convention that the random walk $\alpha = (x_1, \ldots, x_k)$ terminates in the state x_k.

We have also stated earlier that, using the non-analog process Eq. (2.5.17) in which absorption is forbidden, the random variable (2.5.18) is also

---

† This follows as a special case of Theorem 3.5 (see pp. 105ff).

an unbiased estimator of R. This random variable may be written

$$\eta_1(\alpha) = \sum_{m=1}^{k} W_m \frac{\Sigma_r(x_m)}{\Sigma_t(x_m)} \chi_V(x_m), \tag{2.6.4}$$

where W_m is the neutron weight after m collisions, i.e. the product of the probabilities of surviving the first m collisions.

The transport flux, $F(x)$, satisfies the integro-differential equation (see Eq. 2.4.17)

$$\boldsymbol{\omega} \cdot \nabla F(x) + \Sigma_t(x)F(x) = \int \Sigma_t(x')C(x', x)F(x')\,dx' + Q(x). \tag{2.6.5}$$

Let us rewrite Eq. (2.6.5), adding a term to both sides:

$$\boldsymbol{\omega} \cdot \nabla F(x) + [\Sigma_t(x) + \Sigma_{s,\delta}(x)]F(x)$$

$$= \int [\Sigma_t(x')C(x', x) + \Sigma_{s,\delta}(x')\delta(x' - x)]F(x')\,dx' + Q(x)$$

$$= \int [\Sigma_t(x') + \Sigma_{s,\delta}(x')]C^*(x', x)F(x')\,dx' + Q(x). \tag{2.6.6}$$

Here $\delta(x' - x)$ is the Dirac delta function and $C^*(x', x)$ has been redefined so that

$$C^*(x', x) = \frac{\Sigma_t(x')}{\Sigma_t(x') + \Sigma_{s,\delta}(x')} C(x', x) + \frac{\Sigma_{s,\delta}(x')\delta(x' - x)}{\Sigma_t(x') + \Sigma_{s,\delta}(x')}. \tag{2.6.7}$$

The function $\Sigma_{s,\delta}(x)$ may be interpreted as the macroscopic cross section for "scattering" by an artificial isotope. When a scattering collision with this artificial isotope takes place, the particle's position, energy, and direction are left unchanged, so it is as if no collision had taken place.

In addition, a fraction α of the total absorption Σ_a may be associated with "delta events" so that, in effect, the macroscopic cross section identified with the artificial isotope is $\Sigma_{s,\delta} + \alpha\Sigma_a$. We shall treat the absorption associated with delta events in nonanalog fashion. That is, such absorption is not allowed to terminate the history. As is usual, we must then follow weighted histories and enforce a weight reduction by a factor $\Sigma_{s,\delta}/(\Sigma_{s,\delta} + \alpha\Sigma_a)$ when a delta event occurs. With this point of view, we may say that we have associated a macroscopic cross section $\Sigma_s + (1 - \alpha)\Sigma_a$ with "true events." For our purposes in the discussion to follow, this simple dichotomy between true and delta events will suffice. This particular treatment of absorption in delta scattering was suggested in Ref. 8 as an extension of the work of Ref. 9. Later in this section we shall indicate briefly a further means of generalizing the treatment of absorption. Since Eqs. (2.6.5) and (2.6.6) are equivalent, the flux $F(x)$ is invariant under the addition of any amount of this delta function scatterer. The collision density is, however, increased, since the effective total cross section is increased. The idea of delta scattering was suggested by Candelore and Gast, to increase the efficiency of a Monte Carlo resonance escape program, RECAP (Ref. 10). The idea is useful whenever it is desired to increase the collision density without altering the flux, such as is the case in optically thin regions. We shall show, however, that it is not really necessary or even desirable actually to introduce delta events into Monte Carlo simulation. The same benefits may be reaped by utilizing estimators of track length type in combination with estimators of collision type. Our primary purpose in introducing delta scattering here is to provide a convenient theoretical bridge between estimators which record only at collision points and estimators which record continuously along the flight path.

Let

$$\Sigma_t^*(x) = \Sigma_t(x) + \Sigma_{s,\delta}(x) \tag{2.6.8}$$

be the effective total cross section in Eq. (2.6.6). Then, in either the analog or non-analog process of simulating Eq. (2.6.6), the intercollision distance will be foreshortened because of the increase from Σ_t to Σ_t^* (cf. Eq. 2.4.9). The random variable (2.6.3), used to estimate R in the analog simulation of Eq. (2.6.5), becomes

$$\eta_r^*(\alpha) = \sum_{m=1}^{k} \frac{\Sigma_r(x_m)}{\Sigma_t^*(x_m)} \chi_V(x_m) \tag{2.6.9}$$

in the analog simulation of Eq. (2.6.6) since

$$R = \int_\Gamma g(x)\psi(x)\,dx = \int_\Gamma g(x) \frac{\Sigma_t(x)}{\Sigma_t^*(x)} \psi^*(x)\,dx. \tag{2.6.10}$$

Here

$$\psi^*(x) = [\Sigma_t^*(x)/\Sigma_t(x)]\psi(x) \tag{2.6.11}$$

is the collision density associated with Eq. (2.6.6). Similarly, the random variable (2.6.4) becomes

$$\eta_1^*(\alpha) = \sum_{m=1}^{k} W_m^* \frac{\Sigma_t(x_m)}{\Sigma_t^*(x_m)} \chi_V(x_m), \tag{2.6.12}$$

where W_m^* is the survival probability in the problem with delta scattering. Thus,

$$W_m^* = \begin{cases} \dfrac{\Sigma_{s,\delta}}{\Sigma_{s,\delta} + \alpha\Sigma_a} & \text{if collision } m \text{ is a delta event,} \\[2ex] \dfrac{\Sigma_s}{\Sigma_s + (1 - \alpha)\Sigma_a} & \text{if collision } m \text{ is a true event.} \end{cases} \tag{2.6.13}$$

The addition of delta scattering increases the number of collisions per history while the contribution per collision decreases. The random variables (2.6.9) and (2.6.12) are unbiased estimators of R for every choice of $\Sigma_{s,\delta}(x)$, as will be shown in Chapter 3. One may ask what, if any, limit is taken on by η_r^* and η_1^* as the amount of delta scattering tends to infinity in every region? We shall show that these limits exist and give the two estimators mentioned at the beginning of this section.

We have previously established, in our discussion of the transport kernel T (Eqs. 2.4.9 and 2.4.10), that distances between neutron interactions are exponentially distributed. That is, let

$$F(t) = Pr\{S \leq t\} \tag{2.6.14}$$

be the cumulative distribution function for distances S measured in dimensionless units of mean free path (product of total cross section and distance in centimeters). Then $F(t) = 1 - e^{-t}, 0 \leq t \leq \infty$. More generally, it is true that the conditional distance travelled beyond a specific point, given that this point will be passed, has the same distribution. For that distribution is the conditional distribution

$$G(t) = Pr\{S - S_1 \leq t \mid S_1 \leq S\},$$

where S_1 represents the distance from the origin of the path to the point in question. It is easily shown that $G(t) = F(t)$ for any S_1.

Assume that the volume V of phase space may be represented as a product

$$V = V_A \times V_E,$$

where V_A is a homogeneous region of position space and V_E is a subset of energy space. Typically, the set V_E will itself be a set of multiples $E\omega$, $E_{min} \leq E \leq E_{max}$, where ω is any vector on the unit sphere in direction space and (E_{min}, E_{max}) defines the energy interval of interest. For the purposes of the present discussion, all cross sections will be assumed spatially constant in V_A.

Let $x = (\mathbf{r}, \omega, E)$ denote any point of phase space with position component $\mathbf{r}$, unit direction vector ω, and energy E, and let x' be a prospective next collision point for a neutron leaving x. If we write $x' = (\mathbf{r}', \omega, E)$, we have seen that the distance from $\mathbf{r}$ to $\mathbf{r}'$ has an exponential distribution. Let T_x be the random variable representing the distance (mfp) traversed in V_A resulting from the track (x, x'), assuming that this distance is not zero; i.e. assuming that the track intersects V_A. Then T_x has the density function

$$h_x(t) = \begin{cases} e^{-t}, & t < t_{max}, \\ \delta(t - t_{max})e^{-t_{max}}, & t \geq t_{max}, \end{cases} \tag{2.6.15}$$

where t_{max} is the largest possible distance in V_A resulting from a track originating at x.

If D_x denotes the random variable representing distance in centimeters traveled in V_A, assuming this distance is not zero, then D_x has the density

$$k_x(l) = \begin{cases} \Sigma_t e^{-\Sigma_t l}, & l < l_{\max}, \\ \delta(l - l_{\max})e^{-\Sigma_t l_{\max}}, & l \geq l_{\max}, \end{cases} \quad (2.6.16)$$

where Σ_t is the total cross section in V_A at the energy E of the neutron, and $l_{\max}$ is the largest possible value of l given x. Thus, if d is any realization of the random variable D_x, d represents the distance (cm) traveled in V_A by a typical neutron originating at x in the course of simulating Eq. (2.6.5) by analog random walks.

When Eq. (2.6.6) rather than Eq. (2.6.5) is used as a model for the simulation process, the cross section $\Sigma_t^*(x)$ is used to determine intercollision distances. In an analog simulation of Eq. (2.6.6) we shall choose $\alpha = 0$ so that no absorption is associated with the delta isotope. This is done as a matter of convenience to avoid the introduction of weights into the analog process. Thus, the expected number of true events occurring in a length ds is $(\Sigma_s + \Sigma_a)\,ds = \Sigma_t\,ds$. It follows that the density (2.6.16) describes the distribution of distances in V_A between true events in the analog process, with or without delta scattering.

In the non-analog process (Eqs. 2.5.17) a weight reduction must accompany every collision in which absorption might have occurred. If we associate the fraction α of absorption with the delta isotope, then the expected number of true events occurring in a length ds is $\Sigma_{t,\alpha}\,ds$, where $\Sigma_{t,\alpha} \equiv \Sigma_s + (1 - \alpha)\Sigma_a$. Thus, the probability density function for the distance in V_A between true events is

$$g_x(l) = \begin{cases} \Sigma_{t,\alpha} e^{-\Sigma_{t,\alpha} l}, & l < l_{\max}, \\ \delta(l - l_{\max})e^{-\Sigma_{t,\alpha} l}, & l \geq l_{\max}. \end{cases} \quad (2.6.17)$$

We notice that for $\alpha = 0$,

$$g_x(l) = \Sigma_t e^{-\Sigma_t l} \quad \text{for} \quad l < l_{\max},$$

while for $\alpha = 1$,

$$g_x(l) = \Sigma_s e^{-\Sigma_s l} \quad \text{for} \quad l < l_{\max}.$$

As more and more absorption is associated with the delta scattering, the mean free path between true events increases from $1/\Sigma_t$ to $1/\Sigma_s$ because of the increase in weight reduction occurring upon collision with the delta isotope. To summarize our discussion, when simulating Eq. (2.6.6) by the analog process, the density function (2.6.16) is used, while the density function (2.6.17) must be used with the non-analog process.

We examine the analog scheme first. Let d be a sample chosen from the density (2.6.16) so that d represents the length of a typical track originating at x in the analog simulation of Eq. (2.6.5), with a fixed, but arbitrary, choice of $\Sigma_{s,\delta}(x)$. Let $P(n \mid d)$, $n = 0, 1, \ldots$, denote the conditional probability that n pseudo-collisions will occur in the course of travelling a distance d in region V_A. It is not difficult to show (Ref. 9) that $P(n \mid d)$ can be written as an iterated integral which yields the formula

$$P(n \mid d) = \frac{(\Sigma_{s,\delta}d)^n}{n!} e^{-\Sigma_{s,\delta}d}, \qquad d < l_{\max}, \qquad n = 0, 1, \ldots. \quad (2.6.18)$$

Similarly,

$$P(n \mid l_{\max}) = \frac{(\Sigma_{s,\delta}l_{\max})^n}{n!} e^{-\Sigma_{s,\delta}l_{\max}}, \qquad n = 0, 1, \ldots. \quad (2.6.19)$$

Formulas (2.6.18) and (2.6.19) may be combined into the single formula

$$P(n \mid d) = \frac{(\Sigma_{s,\delta}d)^n}{n!} e^{-\Sigma_{s,\delta}d}, \qquad d \leq l_{\max}, \qquad n = 0, 1, \ldots. \quad (2.6.20)$$

Formula (2.6.20) states that the conditional probabilty distribution $P(n \mid d)$ is Poisson in the analog process; it follows that it has mean and variance equal to $\lambda = \Sigma_{s,\delta}d$.

For the non-analog process d must be selected from the density function (2.6.17). Then a similar argument (Ref. 9) reveals that

$$P(n \mid d) = \frac{(\beta d)^n}{n!} e^{-\beta d}, \qquad d \leq l_{\max}, \qquad n = 0, 1, \ldots. \quad (2.6.21)$$

where $\beta = \Sigma_{s,\delta} + \alpha\Sigma_a$. By $P(n \mid d)$ we mean, in this case, the conditional probability that a particle of any initial weight will suffer n pseudo-collisions in the course of traveling a distance d in V_A. Thus, the conditional distribution $P(n \mid d)$ is Poisson for both the analog and non-analog processes.

We now turn our attention to examining the limiting cases of the estimators defined by Eq. (2.6.9) and Eq. (2.6.12). For a given track of length d in region V_A the random variable η_r^* of Eq. (2.6.9) reduces to

$$\eta_r^* \mid d = (n + 1)\frac{\Sigma_r}{\Sigma_t^*} \qquad \text{for } d < l_{\max},$$

where n is the number of pseudo-collisions made along the track, or

$$\eta_r^* \mid d = n\frac{\Sigma_r}{\Sigma_t^*} \qquad \text{for } d = l_{\max}.$$

Given $\epsilon > 0$ and $d < l_{\max}$ we may use Tchebycheff's theorem (Theorem 1.1), applied to an integer k so large that

$$Pr\left\{\left|\eta_r^* \mid d - (\lambda + 1)\frac{\Sigma_r}{\Sigma_t^*}\right| \geq k\frac{\Sigma_r}{\Sigma_t^*}\sqrt{\lambda}\right\} \leq 1/k^2 \leq \epsilon.$$

Thus, with probability $1 - \epsilon$, η_r^* restricted to the track of length d takes on values between

$$(\lambda + 1)\frac{\Sigma_r}{\Sigma_t^*} - k\frac{\Sigma_r}{\Sigma_t^*}\sqrt{\lambda}$$

and

$$(\lambda + 1)\frac{\Sigma_r}{\Sigma_t^*} + k\frac{\Sigma_r}{\Sigma_t^*}\sqrt{\lambda}.$$

Now

$$\lim_{\Sigma_{s,\delta}\to\infty}\left[(\lambda + 1)\frac{\Sigma_r}{\Sigma_t^*} \pm k\frac{\Sigma_r}{\Sigma_t^*}\sqrt{\lambda}\right] = \lim_{\Sigma_{s,\delta}\to\infty}\frac{(\Sigma_{s,\delta}d + 1)\Sigma_r}{\Sigma_t^*} \pm \lim_{\Sigma_{s,\delta}\to\infty}\frac{k\Sigma_r\sqrt{\Sigma_{s,\delta}d}}{\Sigma_t^*}$$

$$= d\Sigma_r,$$

which is the definition of the track length estimator of the reaction rate R, applied to the track of length $d < l_{\max}$ in region V_A. The same conclusion, with d replaced by $l_{\max}$, is valid for $d = l_{\max}$. We have shown that, with probability one, the estimator η_r^* tends to the track length estimator as $\Sigma_{s,\delta}(x)$ tends to infinity in every region.

We have succeeded in proving that the track length estimator arises naturally from estimators of the form of Eq. (2.6.9). Since the track length estimator is a limit of unbiased estimators,† it, too, has been shown to be unbiased. For various choices of $\Sigma_{s,\delta}(x)$, Eq. (2.6.9) defines a family of unbiased estimators of the reaction rate R. The choice $\Sigma_{s,\delta}(x) = 0$ leads to

$$\eta_r^*(\alpha) = \sum_{m=1}^{k} \frac{\Sigma_r(x_m)}{\Sigma_t(x_m)} \chi_V(x_m), \qquad (2.6.22)$$

while the choice $\Sigma_{s,\delta}(x) = \infty$ leads to

$$\eta_r^*(\alpha) = \sum_m \Sigma_{r_m} d_m, \qquad (2.6.23)$$

where d_m is the length of the m-th neutron track in V_A and Σ_{r_m} is the value of Σ_r along the track of length d_m. Explicit problems may be devised for which the variance of (2.6.22) is smaller than that of (2.6.23), and other problems may be devised for which the reverse inequality is true. Generally speaking, the variance of (2.6.22) is smaller in very large and heavily absorbing regions, while (2.6.23) is usually preferred in thin or weakly absorbing regions. Thus, a natural question arises of choosing $\Sigma_{s,\delta}(x)$, for a given problem, in such a way that the variance of (2.6.9) is minimized.

---

† As mentioned earlier, the unbiased nature of collision type estimators will be proved in Chapter 3.

We may also use our knowledge of the distribution (2.6.20) to define new unbiased estimators, obtained by averaging over pseudo-collisions. Thus

$$E[\eta_r^* \mid d] = \sum_{n=0}^{\infty} P(n \mid d)(n+1)\frac{\Sigma_r}{\Sigma_t^*}$$

$$= (1 + \Sigma_{s,\delta}d)\frac{\Sigma_r}{\Sigma_t^*} \qquad \text{if} \qquad d < l_{\max} \qquad (2.6.24)$$

and

$$E[\eta_r^* \mid d] = \Sigma_{s,\delta}l_{\max}\frac{\Sigma_r}{\Sigma_t^*} \qquad \text{if} \qquad d = l_{\max} \qquad (2.6.25)$$

define unbiased estimators of R which should have several advantages over η_r^*. First, one would hope that the variance would be decreased since an averaging process has been performed.[†] Second, in carrying out the Monte Carlo simulation, pseudo-collisions may be ignored in applying Eqs. (2.6.24) and (2.6.25). This will result in significant savings in computer time (compared with actually performing δ-collisions) for large $\Sigma_{s,\delta}$. Third, the numbers $\Sigma_{s,\delta}$ may be regarded in Eqs. (2.6.24) and (2.6.25) as parameters which are used to obtain a variety of simultaneous unbiased estimators of R. It is sometimes possible, through a study of such estimators, to arrive at a practical means of selecting numbers $\Sigma_{s,\delta}$ which approximately minimize the variance.

We notice that Eq. (2.6.24) may be rewritten

$$E[\eta_r^* \mid d] = C_1\frac{\Sigma_r}{\Sigma_t^*} + (1 - C_1)\,d\Sigma_r,$$

where

$$C_1 = \frac{\Sigma_t}{\Sigma_t^*},$$

so that Eq. (2.6.24) amounts to taking a linear combination of the unbiased estimators Σ_r/Σ_t^* and $d\Sigma_r$. The problem of choosing the best (least variance) linear combination of unbiased estimators has been studied in the statistical literature (Ref. 11) and applied to certain random variables arising in Monte Carlo (Refs. 12–15). We shall discuss this point at greater length in Chapter 3. It should be pointed out that $\Sigma_{s,\delta}(x)$ may be made position, energy, and even direction dependent in carrying out this optimization. An indication of how the direction dependence may be effectively used is presented in Ref. 16.

[†] Since the averaging is performed on each collision, rather than over the whole history, this is not necessarily true!

Having completed our analysis of the analog estimator we turn now to the nonanalog estimator η_1^* of Eq. (2.6.12). We notice that in Eq. (2.6.13), the choice

$$\alpha = \frac{\Sigma_{s,\delta}}{\Sigma_{s,\delta} + \Sigma_s}$$

results in the weight reduction factor

$$W_m^* = \frac{\Sigma_{s,\delta} + \Sigma_s}{\Sigma_t^*}$$

on *every* collision. If this choice of α is made, Eq. (2.6.12) becomes considerably simpler, since one need no longer distinguish between true and delta events for the purposes of weight reduction. We therefore make this otherwise arbitrary choice of α and find that, for a given track of length d in V_A, η_1^* reduces to

$$\eta_1^* \mid d = W \frac{\Sigma_r}{\Sigma_a} \left[1 - \left(\frac{\Sigma_s^*}{\Sigma_t^*} \right)^{n+1} \right], \qquad d < l_{max}, \qquad (2.6.26)$$

or

$$\eta_1^* \mid d = W \frac{\Sigma_r}{\Sigma_a} \left[1 - \left(\frac{\Sigma_s^*}{\Sigma_t^*} \right)^{n} \right], \qquad d = l_{max}, \qquad (2.6.27)$$

where n is the number of pseudo-collisions made along the track and W is the neutron weight at the beginning of the track. We use Eq. (2.6.21) to find that, for $d < l_{max}$, $\eta_1^* \mid d$ has mean

$$\mu = W \frac{\Sigma_r}{\Sigma_a} [1 - a e^{-\lambda(a^{-1} - 1)}] \qquad (2.6.28)$$

and standard deviation

$$\sigma = W \frac{\Sigma_r}{\Sigma_a} a \sqrt{e^{-\lambda(a^{-1} - a)} - e^{-2\lambda(a^{-1} - 1)}}, \qquad (2.6.29)$$

where $a = \Sigma_s^*/\Sigma_t^*$, $\lambda = \Sigma_{s,\delta} d$. As before, one may apply Tchebycheff's theorem and evaluate the limits as $\Sigma_{s,\delta} \to \infty$ to find that, with probability one,

$$\lim_{\Sigma_{s,\delta} \to \infty} (\eta_1^* \mid d) = W \frac{\Sigma_r}{\Sigma_a} [1 - e^{-\Sigma_a d}], \qquad d \leq l_{max}. \qquad (2.6.30)$$

This estimator of R is the one suggested by Richtmyer (Ref. 7) for $\Sigma_r = \Sigma_a$ for use in resonance escape calculations. The limit (Eq. 2.6.30) may be interpreted as the original neutron weight less the quantity which would be continuously removed by absorption in travelling a distance d. We notice that, since we have chosen

$$\alpha = \frac{\Sigma_{s,\delta}}{\Sigma_{s,\delta} + \Sigma_s},$$

the quantity α tends to one as $\Sigma_{s,\delta}$ tends to infinity. Thus, from Eq. (2.6.17), the correct density function for distances to be used in conjunction with Richtmyer's estimator is $\Sigma_s e^{-\Sigma_s l}$.

From Eq. (2.6.28) we see that

$$E[\eta_1^* \mid d] = W \frac{\Sigma_r}{\Sigma_a}\left[1 - \frac{\Sigma_s^*}{\Sigma_t^*}\exp\left(\frac{-\Sigma_{s,\delta}\,d\Sigma_a}{\Sigma_s^*}\right)\right], \qquad d < l_{\max}. \quad (2.6.31)$$

Further

$$E[\eta_1^* \mid d] = W \frac{\Sigma_r}{\Sigma_a}\left[1 - \exp\left(\frac{-\Sigma_{s,\delta}l_{\max}\Sigma_a}{\Sigma_s^*}\right)\right], \qquad d = l_{\max}. \quad (2.6.32)$$

These new unbiased estimators of R have the same advantages over (2.6.26) and (2.5.27) discussed for (2.6.24) and (2.6.25). Again the question of choosing optimizing values of $\Sigma_{s,\delta}$ arises and we again defer the discussion of this point to Chapter 3. MacMillan (Ref. 8) has derived formulas like (2.6.31) in which α appears explicitly as a parameter. The interested reader is referred to his paper for details. We remark that it is possible to generalize further the work of this section by introducing a second parameter to appear in all the equations. This parameter would be a fraction representing that portion of Σ_a treated in analog (terminating) fashion. It does not seem worthwhile to develop this theme here, although it could be useful in studying the relative merits of analog and nonanalog treatments of absorption.

2.7 EXTENSIONS OF THE TRACK LENGTH ESTIMATOR

In the previous two sections we have discussed a variety of useful estimators for estimating reaction rates R in transport problems. Chief among these have been three estimators which seem to deserve special attention.

a) The absorption estimator, which records Σ_r/Σ_a upon absorption of the particle.

b) The collision estimator, which records Σ_r/Σ_t upon each particle collision.

c) The track length estimator, which records $d\Sigma_r$ upon each flight resulting in a track of length d.

Of these three basic estimators, it may be said that the third is, in some sense, the most useful. It is useful because it results in low variance in a large variety of regions. In Chapter 3 we shall discuss to a certain extent the possibility touched on briefly in Section 2.6, namely, of taking linear combinations of unbiased estimators to reduce the variance. First, however, we should like to complete our discussion of track length estimators by discussing some extensions and modifications of this basic estimator.

We first describe a modification, applicable to any random walk process in which absorption is permitted, which applies only to the flight immediately preceding absorption. Suppose that, in the normal course of simulating a particle history, it is determined that the particle will be absorbed at point x in region A. Suppose, further, that the particle has just traveled a distance d in A, either from its previous collision point in A or from some point on the boundary of A. The usual track length estimate of a reaction rate R from this last flight is $d\Sigma_r$.

Once it has been decided that absorption will occur in A, use may be made of the fact that the conditional probability density function for l, given that absorption is to occur, is

$$f(l) = \Sigma_t e^{-\Sigma_t l}/(1 - e^{-\Sigma_t D}), \qquad 0 \le l \le D, \qquad (2.7.1)$$

where Σ_t is the total macroscopic cross section in A at the energy of the particle and D is the maximum possible last distance in A. That this is so, whether or not the particle has made its previous collision in A, may easily be seen from the discussion following Eq. (2.6.14).

Now one makes use of Eq. (2.7.1) to obtain the expected last contribution to the track length as

$$E[l\Sigma_r \mid \text{absorption}] = \int_0^D \Sigma_r l f(l)\, dl$$

$$= \frac{\Sigma_r}{\Sigma_t(1 - e^{-\Sigma_t D})} [1 - e^{-\Sigma_t D}(1 + \Sigma_t D)], \quad (2.7.2)$$

the conditional expected value of $l\Sigma_r$, conditioned by absorption in region A.

The averaging performed in defining (2.7.2) removes some of the fluctuation in d on its last flight and should, for this reason, tend to reduce the overall variance. Once again we must distinguish between the variance per collision (which is here rigorously reduced) and the variance per history. We observe that if D is infinite, as in a region A of infinite extent, then

$$E[l\Sigma_r] = \frac{\Sigma_r}{\Sigma_t}, \qquad (2.7.3)$$

the track length estimator reduces to the collision estimator, and the variance on the last flight has been completely eliminated. MacMillan in Ref. 12 has made comparisons of the variance of this modified track length estimator with the unmodified track length estimator for a certain class of idealized problems and has discovered that the modification always reduces variance in such problems.

By making use of knowledge of conditional expected values, as in Eq. (2.7.2), one may develop a variety of new estimators related to the track length estimator, of which Eq. (2.7.2) is one example. This is done as follows (see also Ref. 17).

As before, we adopt the notation $\mathbf{x} = (\mathbf{r}, \mathbf{E})$, where $\mathbf{r}$ is a vector of position space and $\mathbf{E}$ a vector of velocity space. Let $\boldsymbol{\omega}$ denote the unit vector

$E/\|E\|$. Suppose that a neutron with velocity E appears in a region A in which all cross sections are independent of position. The neutron may, for example, have just been born in A. Or it may have collided in A, thus acquiring the velocity E. Finally, it may have just entered A from a region contiguous to it. Let D be the distance along ω from the position r of the neutron to the nearest point of the boundary of A. Then D is the largest possible distance the neutron can travel in A before leaving A. Now let

$$\Sigma_t(\mathbf{r}, E) = \Sigma_a(\mathbf{r}, E) + \Sigma_s(\mathbf{r}, E),$$

where Σ_a and Σ_s are, respectively, the absorption and scattering cross sections in A at velocity E. For convenience, we omit the arguments $\mathbf{r}$, E and write the identity

$$e^{-\Sigma_t D} + \frac{\Sigma_a}{\Sigma_t}(1 - e^{-\Sigma_t D}) + \frac{\Sigma_s}{\Sigma_t}(1 - e^{-\Sigma_t D}) = 1. \qquad (2.7.4)$$

We note that the first term represents the probability that the neutron will pass directly through A without suffering a collision, while the second and third terms represent, respectively, the probabilities of absorption and scattering in A on the next collision. If $\mathbf{r}'$ denotes the position of the next collision of the neutron, then the distance l from $\mathbf{r}$ to $\mathbf{r}'$ is a random variable whose density function is

$$f(l) = \Sigma_t(\mathbf{r}') \exp\left[-\int_0^{l=\|\mathbf{r}-\mathbf{r}'\|} \Sigma_t(\mathbf{r} + s\omega)\, ds \right], \qquad 0 \le l \le \infty. \quad (2.7.5)$$

On the other hand, if l^* is the random variable representing the distance in A resulting in the track from $\mathbf{r}$ to $\mathbf{r}'$, then l^* has the density function

$$g(l^*) = \begin{cases} \Sigma_t e^{-\Sigma_t l^*}, & l^* < D, \\ \delta(l^* - D)e^{-\Sigma_t D}, & l^* \ge D. \end{cases} \qquad (2.7.6)$$

Even if $\mathbf{r}$ is outside A, one can compute the density of l^* provided the ray $\mathbf{r} + s\omega$, $s \ge 0$, intersects A. The conditional density then is

$$g_1(l^*) = \begin{cases} \delta(l^*)(1 - e^{-T_0}), & l^* = 0, \\ e^{-T_0}e^{-l^*}, & 0 < l^* < D_{\max}, \\ e^{-T_0}\delta(l^* - D_{\max})e^{-D_{\max}}, & l^* = D_{\max}, \end{cases} \qquad (2.7.7)$$

where l^* is in units of mean free path, T_0 is the distance (in mfp) from $\mathbf{r}$ to A along ω, and $D_{\max}$ is the largest distance in A (in mfp) the neutron may travel before leaving A. We shall work with the function (2.7.6) for convenience; the generalization involving Eq. (2.7.7) follows directly from the same arguments.

Now one may also derive the conditional density function for the random variable which represents the distance traveled in A conditioned by any subset of the three events:

a) direct passage through A without collision,
b) absorption in A on the next collision,
c) scattering in A on the next collision.

Averaging over these various conditional density functions will result in new extimators of R which may then be compared on the basis of variance and computing time.

To derive the new estimators, we note that the conditional density function for l conditioned by (a) is

$$f_a(l) = \delta(l - D), \qquad 0 \leq l \leq D \tag{2.7.8}$$

and the conditional density function for l conditioned by either (b) or (c) is

$$f_b(l) = f_c(l) = \frac{\Sigma_t e^{-\Sigma_t l}}{1 - e^{-\Sigma_t D}}, \qquad 0 \leq l \leq D. \tag{2.7.9}$$

Now, if S is any subset of the events (a), (b), (c), the conditional density function of l given S is simply

$$f_S(l) = \frac{\sum_{x \in S} f_x(l) P_x}{\sum_{x \in S} P_x}, \qquad 0 \leq l \leq D, \tag{2.7.10}$$

where P_x is the probability of the event (x) and f_x is defined by either Eq. (2.7.8) or Eq. (2.7.9). The conditional expected value, E_S, is then defined in the usual way as

$$E_S = \Sigma_r \int_0^D l f_S(l) \, dl.$$

The resulting random variable, E_S, is used as the estimate of reaction rate if any of the events in S occurs. It is also possible to condition separately for subsets S_1, S_2 of the conditions (a), (b), (c).

To illustrate the general derivation, suppose $S = \{(a), (b), (c)\}$, so that it is desired to condition for all three possible outcomes. Then

$$E_{abc} = \frac{\Sigma_r \int_0^D l \left\{ e^{-\Sigma_t D}\delta(l - D) + \dfrac{\Sigma_a}{\Sigma_t}(1 - e^{-\Sigma_t D})\dfrac{\Sigma_t e^{-\Sigma_t l}}{(1 - e^{-\Sigma_t D})} + \dfrac{\Sigma_s}{\Sigma_t}(1 - e^{-\Sigma_t D})\dfrac{\Sigma_t e^{-\Sigma_t l}}{(1 - e^{-\Sigma_t D})} \right\} dl}{e^{-\Sigma_t D} + \dfrac{\Sigma_a}{\Sigma_t}(1 - e^{-\Sigma_t D}) + \dfrac{\Sigma_s}{\Sigma_t}(1 - e^{-\Sigma_t D})}$$

$$= \frac{\Sigma_r}{\Sigma_t}(1 - e^{-\Sigma_t D}).$$

In Table 2.1 we have displayed all of the random variables which may be derived by the above considerations. We use the following notation: d denotes the actual distance traveled in A and

$$E_b = E_c = \frac{\Sigma_r}{\Sigma_t} \frac{[1 - e^{-\Sigma_t D}(\Sigma_t D + 1)]}{(1 - e^{-\Sigma_t D})},$$

$$E_{ab} = \Sigma_r \left\{ \frac{\Sigma_s D e^{-\Sigma_t D} + \frac{\Sigma_a}{\Sigma_t}(1 - e^{-\Sigma_t D})}{\Sigma_s e^{-\Sigma_t D} + \Sigma_a} \right\},$$

$$E_{ac} = \Sigma_r \left\{ \frac{\Sigma_a D e^{-\Sigma_t D} + \frac{\Sigma_s}{\Sigma_t}(1 - e^{-\Sigma_t D})}{\Sigma_a e^{-\Sigma_t D} + \Sigma_s} \right\},$$

$$E_{abc} = \frac{\Sigma_r}{\Sigma_t}(1 - e^{-\Sigma_t D}).$$

In Table 2.1 we exhibit the subsets S_1, S_2 which are being conditioned separately, together with the random variable which defines the estimator upon the actual occurrence of condition (a), (b), or (c). For any conditions which fail to be included in $S_1 \cup S_2$ the track length estimator is used. We notice that random variable I is merely the track length estimator, since conditioning for (a) separately results in no change in the random variable when condition (a) occurs, while random variable II is the modified track length estimator discussed earlier see Eq. 2.7.2. A dash in Table 2.1 denotes the empty set.

TABLE 2.1

TABLE OF RANDOM VARIABLES

	S_1	S_2	Estimator if (a)	Estimator if (b)	Estimator if (c)
I	a	—	$\Sigma_r D$	$\Sigma_r d$	$\Sigma_r d$
II	a	b	$\Sigma_r D$	E_b	$\Sigma_r d$
III	a	c	$\Sigma_r D$	$\Sigma_r d$	E_c
IV	a, b	—	E_{ab}	E_{ab}	$\Sigma_r d$
V	a, b	c	E_{ab}	E_{ab}	E_c
VI	a, c	—	E_{ac}	$\Sigma_r d$	E_{ac}
VII	a, c	b	E_{ac}	E_b	E_{ac}
VIII	b, c	a	$\Sigma_r D$	E_b	E_c
IX	a, b, c	—	E_{abc}	E_{abc}	E_{abc}

When the density function (2.7.7) is used to derive estimators for points $\mathbf{r}$ outside of A, the result is merely to multiply the values in the table by the attenuation factor e^{-T_0}. We shall derive one of these estimators from a different point of view later (see Eq. 3.6.18). Such estimators would prove extremely useful in cases where few particles actually reached the region A and estimates of the reaction rate R are obtainable by virtue of particles heading in directions which intersect A. Such is the case, for example, in one-dimensional deep penetration problems.

As for the relative merits of the estimators of Table 2.1, the efficiencies of the various estimators depend on the characteristics of the region to which the estimators are applied. Experimentation has been performed on multigroup thermal problems which reveals that estimators II, IV, and IX are capable of reducing the variance (compared with I), but with some increase in computation time per history. Taking into account MacMillan's analysis and experimentation on more practical problems, it seems difficult to beat II in overall efficiency. However, since such comparisons are obviously computer dependent, it is worth noting that all of the estimators of Table 2.1 are unbiased and by making the estimator region dependent, it is possible that rather large increases in efficiency may result.

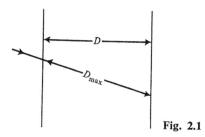

Fig. 2.1

We have previously commented on the distinction between variance per collision and variance per history. It is the latter quantity which determines the efficiency of a given estimator. In this connection we should like to cite the following example in which the track length estimator (estimator I) has zero variance but the estimator IX has positive variance. Consider a slab of normal thickness D cm (see Fig. 2.1) in which only delta scattering is permitted. Suppose a monodirectional stream of neutrons impinges on the slab. Clearly, the track length estimate of the reaction rate is $\Sigma_r D_{max}$ for every neutron, where D_{max} is distance traversed by every neutron. But the variance of IX certainly depends on the number of collisions made within the slab. This example seems to preclude a theoretical ranking of the estimators according to variance.

REFERENCES

1. G. E. FORSYTHE and R. A. LEIBLER, "Matrix Inversion by a Monte Carlo Method," *Math. Tables Aids to Comp.*, **4**, 127 (1950).

2. W. WASOW, "A Note on the Inversion of Matrices by Random Walks," *Math. Tables Aids to Comp.*, **6**, 78 (1952).

3. R. S. VARGA, *Matrix Iterative Analysis*, Prentice-Hall, Englewood Cliffs, New Jersey (1962).

4. G. GOERTZEL and M. H. KALOS, "Monte Carlo Methods in Transport Problems," *Progress in Nuclear Energy*, Vol. II, Series I, Physics and Mathematics, ed. D. J. Hughes, J. E. Sanders, J. Horowitz, Pergamon Press, New York (1958), pp. 315–369.

5. J. SPANIER, "Monte Carlo Methods and Their Application to Neutron Transport Problems," *Bettis Atomic Power Laboratory, WAPD*-195 (July, 1959).

6. M. H. KALOS, "On the Estimation of Flux at a Point by Monte Carlo," *Nucl. Sci. Eng.*, **16**, 111 (1963).

7. R. D. RICHTMYER, "The Monte Carlo Calculation of Resonance Capture in Reactor Lattices," *United Nations International Conference on Peaceful Uses of Atomic Energy, 2nd, Geneva, 1952, Proceedings*.

8. D. B. MACMILLAN, "Note on a Paper by J. Spanier Concerning Monte Carlo Estimators," *J. SIAM Appl. Math.*, **15**, 264 (1967).

9. J. SPANIER, "Two Pairs of Families of Estimators for Transport Problems," *J. SIAM Appl. Math.*, **14**, 702 (1966).

10. N. R. CANDELORE and R. C. GAST, "RECAP-2 A Monte Carlo Program for Estimating Epithermal Capture Rates in Rod Arrays," *Bettis Atomic Power Laboratory Report WAPD-TM*-427 (1964).

11. M. HALPERIN, "Almost Linearly Optimum Combination of Unbiased Estimates," *J. Am. Stat. Assoc.*, **56**, 36 (1961).

12. D. B. MACMILLAN, "Comparison of Statistical Estimators for Neutron Monte Carlo Calculations," *Nucl. Sci. Eng.*, **26**, 366 (1966).

13. G. L. BURROWS and D. B. MACMILLAN, "Confidence Limits for Monte Carlo Calculations," *Nucl. Sci. Eng.* **22**, 384 (1965).

14. M. A. MARTINO and W. W. STONE, "TRAM, A Monte Carlo Thermal Neutron Code for the IBM 704," *Knolls Atomic Power Laboratory Report KAPL*-2039 (1959).

15. E. M. GELBARD, H. B. ONDIS, and J. SPANIER, "MARC-A Multigroup Monte Carlo Program for the Calculation of Capture Probabilities," *Bettis Atomic Power Laboratory Report WAPD-TM*-273 (1962).

16. N. M. STEEN, "A Simple Method to Improve the Efficiency of the Σ_a/Σ_t Estimator in Certain Monte Carlo Programs," *Bettis Atomic Power Laboratory Report WAPD-TM*-609 (Oct., 1966).

17. E. M. GELBARD, L. A. ONDIS II, and J. SPANIER, "A New Class of Monte Carlo Estimators," *J. SIAM Appl. Math.*, **14**, 697 (1966).

3

Standard Variance Reduction Techniques

3.1 INTRODUCTION

In Chapters 1 and 2 we have discussed many Monte Carlo methods which play a key role in the solution of particle transport problems. We have exhibited a variety of ways to sample from distributions and have displayed random walk processes which may be used to simulate particle behavior. In addition, we have presented the basic estimators of transport theory and have shown how they may be used to estimate quantities of interest.

Naturally, the efficiency of an estimator is roughly measured by its variance; generally, the smaller the variance for a fixed number of histories, the more efficient is the estimator. However, since the methods we discuss will normally be applied to a digital computer, the efficiency for a given problem must also take into account the running time per history. Hammersley and Handscomb (Ref. 1) define the efficiency of a Monte Carlo process as a number which is inversely proportional to the product of the variance and the labor expended in obtaining the estimate. This definition is also appropriate for our purposes and in this chapter we shall focus attention on standard methods for reducing the variance of the sampling process, keeping in mind that decreases in variance must be balanced against possible increases in cost per history. We use the term "standard" here to distinguish the methods from the special variance reduction devices, based on superposition and reciprocity, whose applications will be discussed in greater detail in Chapters 4–6. We should also remark that in our choice of standard methods we have been prejudiced in favor of those methods which we have used most frequently and successfully in our own applications. Among the important methods not discussed here is the method of conditional Monte Carlo, discovered by Trotter and Tukey (Ref. 2) and applied to the transport equation by Drawbaugh in Ref. 3. Hammersley and Handscomb (Ref. 1) have devoted a chapter to this potentially very useful method, and the interested reader is also referred to Refs. 4 and 5. Another general method for estimating definite integrals is based on the use of orthogonal polynomials and seems first to have been mentioned by Ermakov and Zolotukhin (Ref. 6). Reference 1 also discusses their method, as does Handscomb in Ref. 7. The use of the exponential transformation has been discussed at length in Refs. 8 and 9. As is pointed out in Ref. 10 and again in Section 3.7, it may be regarded as a special kind of importance sampling within the very general framework of our discussion in Section 3.7.

One further point about efficiency seems worth mentioning. The running time for a particular calculation is strongly dependent on the specific computer on which the calculation is performed. With the rapid advances being made in the design and performance of high-speed digital equipment, statements made today about the time involved in a certain calculation may be altered drastically by the computer of tomorrow. Thus, we take the point of view that we shall focus our attention primarily on reductions in variance but mention, wherever possible, the effect of a given variance reduction device on running time.

3.2 GENERAL PRINCIPLES

Every Monte Carlo calculation can be regarded as yielding an estimate of one or more definite integrals, namely, the expected values of the estimating random variables. This statement, while obvious, hides the slightly less obvious point that in estimating an expected value, we have replaced the original problem derived from a physical model by an equivalent one phrased in terms of a probability model.

To be more specific, initially the transport problem to be solved may be formulated in terms of estimating the value of an integral

$$I = \int_\Gamma g\psi, \tag{3.2.1}$$

where Γ is phase space, g is a known function† and ψ satisfies the transport equation, which we write formally as

$$L\psi = S. \tag{3.2.2}$$

This problem is then replaced by the problem of estimating the expected value

$$I = \int_\Omega \xi(C) \, d\mu(C), \tag{3.2.3}$$

where Ω is the space consisting of all random walk histories, ξ is an unbiased estimator (on Ω) of I, and $\mu(C)$ is a probability function defined on Ω. As we

---

† Initially, at least, we shall not prejudice ourselves against the possibility that g is a Dirac delta function, in which case the integral (3.2.1) is the value of ψ at a point.

have already seen, the unknown expected value is invariably estimated by the average

$$\bar{\xi}_N = \frac{1}{N} \sum_{i=1}^{N} \xi(C_i) \qquad (3.2.4)$$

of the values of ξ on N histories, i.e. the sample mean of ξ.

In Chapter 2 we have shown, in several instances, how this equivalence between physical and probability models comes about. The construction of the probability function $\mu(C)$ in complete detail is beyond the scope of the present book; it is done in general in Ref. 11. We should, nevertheless, like to give some feeling for how the probability function $\mu(C)$ is constructed from the underlying random walk process. This can be done without dwelling on the measure-theoretic aspects of the construction and should prove useful, indeed really necessary, to the understanding of some of the variance reduction methods we discuss in this chapter. Accordingly, we shall first extend and generalize some of the concepts introduced in Chapter 2 for continuous random walk processes.

3.3 RANDOM WALK PROCESSES AND THEIR ASSOCIATED PROBABILITIES

In this section we shall derive a number of important and fundamental results which establish a rigorous connection between the physical model, based on the transport equation, and the probability model used in the construction of the Monte Carlo sampling. Primarily we shall show that if the physical model is subcritical, then so is the probability model corresponding to a direct simulation of the physical model. By a subcritical probability model we mean one in which the number of collisions made to termination is finite with probability one. Later we discuss extensions of these results to nonanalog simulation processes. Although the results proved in this section are used time and again in proofs in later sections, those readers interested mainly in the estimators of transport theory or the applications may want to omit the material of Section 3.3 beginning with Lemma 3.1, at least on first reading. If the reader is willing to ignore the possibility that particles may make an infinite number of collisions and to assume that Neumann series encountered later in proofs of theorems always converge, then the indicated material may be skipped without loss of continuity in the exposition. Nevertheless, because of its basic importance, the material of Section 3.3 is useful in understanding the mathematical point of view we have adopted in an attempt to make our discussion of Monte Carlo methods rigorous and self-contained.

We have loosely defined the space Ω as the space of all possible random walks in phase space. Now we shall be more specific and define Ω to be the space of all chains† $C = \{\alpha_1, \alpha_2, \ldots\}$. Here α_i is a pair (P_i, τ) or (P_i, τ'), P_i being a point of the physical phase space Γ, and τ and τ' designating states which correspond to nontermination and termination, respectively, of the random walk. Intuitively, the typical chain

$$C = \{(P_1, \tau), (P_2, \tau), (P_3, \tau), (P_4, \tau'), (P_4, \tau'), \ldots\}$$

denotes a random walk originating at point P_1 of phase space, passing next to the state P_2, then to P_3, and finally to P_4, where it terminates.‡ As usual, the P_i are to be identified with collision points. It is natural to define the *length* $n(C)$ of the chain C as the smallest integer k such that the element α_k of C has the form $\alpha_k = (P_k, \tau')$. The function $n(C)$ on Ω represents the number of collisions made to termination of the random walk chain C.

Let $\Lambda_k = \{C \in \Omega \mid n(C) = k\}$, $k = 1, 2, \ldots$, be the subset of Ω consisting of those chains of length k and let $\Lambda_\infty = \{C \in \Omega \mid n(C) = \infty\}$ be the chains of infinite length. It is evident that the sets Λ_k for k finite and Λ_∞ are pair-wise disjoint and together make up the space Ω. That is,

$$\Omega = \left(\bigcup_{k=1}^{\infty} \Lambda_k\right) \cup \Lambda_\infty \tag{3.3.1}$$

and $\Lambda_i \cap \Lambda_j = \emptyset, i \neq j$, so that the sets $\Lambda_k, \Lambda_\infty$ form a decomposition of the space Ω.

The probability function μ is a real-valued function defined on certain subsets of Ω, which is to be interpreted as the probability of occurrence of the subset. In our case, μ will always be defined by an integral. We shall now show how μ is defined on the Λ_i in terms of certain sequences of functions which serve to characterize the random walk process.

Let Γ denote the physical phase space and Γ^n the space of all *n*-tuples of points of Γ. By a *random walk process*§ $\{f_n, p_n\}$ we mean a sequence $f_n(P_1, \ldots, P_n)$ of continuous probability density functions defined on Γ^n together with a sequence $p_n(P_1, \ldots, P_n)$ of functions on Γ^n with the properties‖

$$F_n(P_1, \ldots, P_k, \infty, \ldots, \infty) = F_k(P_1, \ldots, P_k) \qquad (k < n, n \geq 1),$$
$$\tag{3.3.2}$$

where

$$F_n(P_1, \ldots, P_n) = \int_{-\infty}^{P} \cdots \int_{-\infty}^{P} f_n(P_1', \ldots, P_n') \, dP_1' \cdots dP_n' \tag{3.3.3}$$

---

† In this context a chain is merely an infinite sequence.

‡ In Chapter 2 such a chain would have been denoted $C = (P_1, P_2, P_3, P_4)$. It is mathematically more convenient to let Ω be a space of infinite sequences, as we do here.

§ See first footnote on p. 92.

‖ The symbol ∞ denotes the point of Γ each of whose components is infinite.

is the distribution function of f_n and

$$0 \leq p_n(P_1, \ldots, P_n) \leq 1, \qquad n = 1, 2, \ldots \tag{3.3.4}$$

for all $(P_1, \ldots, P_n)$. Intuitively, $f_n(P_1, \ldots, P_n)$ represents the probability density of a random walk chain involving the ordered sequence of collision points $(P_1, \ldots, P_n)$ and $p_n(P_1, \ldots, P_n)$ is the probability of terminating the random walk at P_n. Then $q_n = 1 - p_n$ is the probability of continuing the random walk beyond the state P_n.

This notion of a random walk process and the basic construction of the probability function $\mu(C)$ are due to Albert (see Ref. 12). The reader should realize that the definition given here of a random walk process generalizes the specific constructions for discrete and continuous cases given in Chapter 2. We shall usually not have need for the generality included here; for example, $p_n(P_1, \ldots, P_n)$ will most often be independent of $P_1, \ldots, P_{n-1}$ and of n; for those familar with the term, we shall be dealing almost exclusively with Markov processes.† Nonetheless, it seems worth pointing out that non-Markov processes may be treated as well.

The probability function μ is derived from the random walk process $\{f_n, p_n\}$ making use of a rather technical theorem of Tulcea (Ref. 14). For our purposes it will suffice to observe that

$$\mu(\Lambda_k) = \int_\Gamma \cdots \int_\Gamma f_k(P_1, \ldots, P_k) \prod_{i=1}^{k-1} q_i(P_1, \ldots, P_i) p_k(P_1, \ldots, P_k) \, dP_1 \cdots dP_k, \tag{3.3.5}$$

so that the multiple integral on the right is the probability of a chain of length exactly k. More generally, for any random variable ξ, the expected value of ξ taken over Λ_k is

$$\int_{\Lambda_k} \xi \, d\mu = \int_\Gamma \cdots \int_\Gamma \xi(P_1, \ldots, P_k) f_k(P_1, \ldots, P_k)$$
$$\times \prod_{i=1}^{k-1} q_i(P_1, \ldots, P_i) p_k(P_1, \ldots, P_k) \, dP_1 \cdots dP_k.$$

This is an intuitively plausible definition, because in the integral f_k gives the probability density of the chain $(P_1, \ldots, P_k)$, $\prod_{i=1}^{k-1} q_i$ guarantees that the walk does not terminate at $P_1, \ldots, P_{k-1}$, and p_k assures the termination at P_k. Thus, the probability function arises as a natural consequence of the random walk process and has a similar definition on other subsets of Ω with which a probability may be associated. In Eq. (3.3.5) and in all work that follows we adopt the convention that the product $\prod_{i=1}^{0} = 1$.

From a practical point of view, it is necessary to satisfy conditions on the random walk process which guarantee that the length $n(C)$ is finite with

---

† See, for example, Ref. 13 for a definition and discussion of Markov processes.

probability one, i.e. that

$$\sum_{k=1}^{\infty} \mu(\Lambda_k) = 1,$$

or, equivalently, that

$$\mu(\Lambda_\infty) = 0. \tag{3.3.6}$$

When condition (3.3.6) is met we shall say that the probability model is *subcritical*. In Chapter 2 we discussed briefly how this condition came about in a more or less natural way for discrete random walk processes. Now we should like to state conditions under which it will be guaranteed for continuous random walk processes. For this we shall need some definitions and terminology.

Based on the integral equation

$$\psi(P) = \int_\Gamma K(P, P')\psi(P') \, dP' + S(P) \tag{3.3.7}$$

for the collision density $\psi(P)$, an analog random walk process for a (possibly multiplying) medium is defined by the choices

$$f_1(P) = S(P), \tag{3.3.8}$$

$$f_n(P_1, \ldots, P_n) = \prod_{l=2}^{n} \left[\frac{K(P_l, P_{l-1})}{\int_\Gamma K(P_l, P_{l-1}) \, dP_l} \right] S(P_1), \qquad n \geq 2, \tag{3.3.9}$$

and

$$p_n(P_1, \ldots, P_n) = p(P_n) = \Sigma_a(P_n)/\Sigma_t(P_n), \tag{3.3.10}$$

where Σ_a and Σ_t are, respectively, the absorption and total macroscopic cross sections. In a non-multiplying medium one has (see footnote, p. 59)

$$\int_\Gamma K(Q, P) \, dQ = \Sigma_s(P)/\Sigma_t(P) = 1 - p(P) = q(P) \leq 1.$$

In a multiplying medium one has, in any event,

$$q(P) \leq \int_\Gamma K(Q, P) \, dQ \equiv c(P), \tag{3.3.11}$$

with equality holding at all points P where no neutron multiplication occurs. The quantity $c(P)$ is the mean number of secondaries at P.

The analog random walk process may also be defined in an exactly analogous way, based on the integral equation (2.4.20) for the density function $\chi(\mathbf{r}, \mathbf{E})$. We take

$$f_1(P) = Q(P), \tag{3.3.8a}$$

$$f_n(P_1, \ldots, P_n) = \prod_{l=2}^{n} \left[\frac{L(P_l, P_{l-1})}{\int_\Gamma L(P_l, P_{l-1}) \, dP_l} \right] Q(P_1), \qquad n \geq 2, \tag{3.3.9a}$$

and

$$p_n(P_1, \ldots, P_n) = p(P_n). \tag{3.3.10a}$$

We recall that

$$K(\mathbf{r}, \mathbf{E}; \mathbf{r}', \mathbf{E}') = C(\mathbf{E}', \mathbf{E}; \mathbf{r}')T(\mathbf{r}', \mathbf{r}; \mathbf{E}),$$

$$L(\mathbf{r}, \mathbf{E}; \mathbf{r}', \mathbf{E}') = T(\mathbf{r}', \mathbf{r}; \mathbf{E}')C(\mathbf{E}', \mathbf{E}; \mathbf{r}).$$

We notice that the kernel K first alters the energy from $\mathbf{E}'$ to $\mathbf{E}$ at point $\mathbf{r}'$ (by means of a collision) and then moves the particle from $\mathbf{r}'$ to $\mathbf{r}$ at the new energy $\mathbf{E}$. The kernel L, on the other hand, first moves the particle from $\mathbf{r}'$ to $\mathbf{r}$ at energy $\mathbf{E}'$ and then alters the energy from $\mathbf{E}'$ to $\mathbf{E}$ at position $\mathbf{r}$. If we let $c'(P) = \int L(Q, P)\, dQ$ be the mean number of secondaries at P via the process described by the kernel L, then, in general, $c(P) \neq c'(P)$. In fact,

$$c'(\mathbf{r}', \mathbf{E}') = \int T(\mathbf{r}', \mathbf{r}; \mathbf{E}')c(\mathbf{r}, \mathbf{E}')\, d\mathbf{r},$$

so that the mean number of secondaries resulting from the use of the kernel L is a kind of spatial average of the mean number resulting from the use of K. Yet the same physical processes underlie Eqs. (2.4.20) and (3.3.7). For this reason one might expect that analog random walks based on these equations would be identical. Indeed, despite the formal differences between the two equations, analog processes characterized by (3.3.8)–(3.3.10) and by (3.3.8a)–(3.3.10a) are usually identical in their execution. As we shall see later in Chapter 3, the densities ψ and χ are both constructed, generally, in one and the same analog simulation. If we sometimes choose to work with (2.4.20) and prefer Eq. (3.3.7) at other times, this is simply a matter of convenience.

From the properties of the probability function μ, we have

$$1 = \mu(\Omega) = \sum_{k=1}^{\infty} \mu(\Lambda_k) + \mu(\Lambda_\infty),$$

so that

$$\mu(\Lambda_\infty) = 1 - \sum_{k=1}^{\infty} \mu(\Lambda_k)$$

$$= 1 - \sum_{k=1}^{\infty} \int_\Gamma \cdots \int_\Gamma f_k(P_1, \ldots, P_k) \prod_{i=1}^{k-1} q(P_i)p(P_k)\, dP_1 \cdots dP_k$$

from (3.3.5) and assuming $q_i(P_1, \ldots, P_i) = q(P_i)$ for convenience.

In order to display a formula for $\mu(\Lambda_\infty)$ for a general random walk process $\{f_n, p_n\}$ we make use of the following lemma.

Lemma 3.1 For every $N \geq 1$,

$$\sum_{k=1}^{N} \int_\Gamma \cdots \int_\Gamma f_k(P_1, \ldots, P_k) \prod_{i=1}^{k-1} q(P_i)p(P_k)\, dP_1 \cdots dP_k$$

$$= 1 - \int_\Gamma \cdots \int_\Gamma f_N(P_1, \ldots, P_N) \prod_{i=1}^{N} q_i\, dP_1 \cdots dP_N.$$

Proof. Since $p(P_k) = 1 - q(P_k)$,

$$\sum_{k=1}^{N} \int_{\Gamma} \cdots \int_{\Gamma} f_k \prod_{i=1}^{k-1} q(P_i) p(P_k) \, dP_1 \cdots dP_k$$

$$= \sum_{k=1}^{N} \int_{\Gamma} \cdots \int_{\Gamma} f_k \prod_{i=1}^{k-1} q(P_i) \, dP_1 \cdots dP_k - \sum_{k=1}^{N} \int_{\Gamma} \cdots \int_{\Gamma} f_k \prod_{i=1}^{k} q(P_i) \, dP_1 \cdots dP_k$$

$$= \int_{\Gamma} f_1(P) \, dP + \sum_{k=2}^{N} \int_{\Gamma} \cdots \int_{\Gamma} f_{k-1}(P_1, \ldots, P_{k-1}) \prod_{i=1}^{k-1} q(P_i) \, dP_1 \cdots dP_{k-1}$$

$$- \sum_{k=1}^{N} \int_{\Gamma} \cdots \int_{\Gamma} f_k(P_1, \ldots, P_k) \prod_{i=1}^{k} q(P_i) \, dP_1 \cdots dP_k$$

$$= 1 - \int_{\Gamma} \cdots \int_{\Gamma} f_N(P_1, \ldots, P_N) \prod_{i=2}^{N} q(P_i) \, dP_1 \cdots dP_N.$$

Note that Lemma 3.1 has a simple physical interpretation. The left-hand side of the equality is the probability of absorption on one of the first N collisions, while the integral term on the right-hand side is the probability of surviving N collisions.

From Lemma 3.1 we see that

$$\mu(\Lambda_\infty) = \lim_{N \to \infty} \int_{\Gamma} \cdots \int_{\Gamma} f_N(P_1, \ldots, P_N) \prod_{i=1}^{N} q(P_i) \, dP_1 \cdots dP_N, \quad (3.3.12)$$

For the analog process defined by Eqs. (3.3.8), (3.3.9), and (3.3.10) we have

$$\mu(\Lambda_\infty) = \lim_{N \to \infty} \int_{\Gamma} \cdots \int_{\Gamma} \frac{K(P_N, P_{N-1})}{c(P_{N-1})} \cdots \frac{K(P_2, P_1)}{c(P_1)} S(P_1) \prod_{i=1}^{N} q(P_i) \, dP_1 \cdots dP_N$$

$$\leq \lim_{N \to \infty} \int_{\Gamma} \cdots \int_{\Gamma} K(P_N, P_{N-1}) \cdots K(P_2, P_1) S(P_1) \, dP_1 \cdots dP_N \quad (3.3.13)$$

by (3.3.11) and the fact that $\sup_{P \in \Gamma} q(P) \leq 1$.

We should now like to state conditions on the kernel K which guarantee that the probability model based on the analog random walk process is subcritical. The following notation will be convenient. Define

$$K_n(P, P_1) = \int_{\Gamma} \cdots \int_{\Gamma} K(P, P_n) K(P_n, P_{n-1}) \cdots K(P_2, P_1) \, dP_2 \cdots dP_n,$$

$$n > 1, \quad (3.3.14)$$

where

$$K_1(P, P_1) = K(P, P_1) \quad (3.3.15)$$

and define integral operators $\mathscr{K}_n$ by

$$\mathscr{K}_n f(P) = \int_{\Gamma} K_n(P, P_1) f(P_1) \, dP_1, \quad n \geq 1. \quad (3.3.16)$$

We note that $\mathscr{K}_n = \mathscr{K}_1^n$ and that the Neumann series corresponding to Eq. (3.3.7) is

$$\mathscr{K}S + \mathscr{K}_1S + \mathscr{K}_2S + \cdots.$$

It is also evident that

$$K_{n+m}(P, P_1) = \int_\Gamma K_n(P, Q)K_m(Q, P_1)\, dQ \tag{3.3.17}$$

for all $m, n \geq 1$.

We shall assume the following two conditions† concerning the kernel K:

a) There exists a constant $M < \infty$ such that

$$\sup_{P' \in \Gamma} \int_\Gamma K(P, P')\, dP \leq M. \tag{3.3.18}$$

b) There exists a constant $c < 1$ and an integer N such that for all integers $n \geq N$, the iterated integral

$$\int_\Gamma K_n(P_n, P)\, dP_n = \int_\Gamma K(P_n, P_{n-1}) \cdots K(P_1, P)\, dP_1 \cdots dP_n \leq c < 1 \tag{3.3.19}$$

for all $P \in \Gamma$.

Both conditions (a) and (b) appear to be reasonable ones. For example, in a nonmultiplying medium we have seen that

$$\int_\Gamma K(Q, P)\, dQ = \Sigma_s(P)/\Sigma_t(P) \leq 1 \qquad \text{for all } P \in \Gamma$$

so that (a) is satisfied with $M = 1$. In such a case, if there is a subregion of Γ for which $\Sigma_s(P) < \Sigma_t(P)$ (an absorbing subregion) and if the differential scattering cross section at all P is nonzero for all scattering directions, then it may be shown that

$$\iint_{\Gamma\Gamma} K(P_2, P_1)K(P_1, P)\, dP_1\, dP_2 < 1 \tag{3.3.20}$$

for all P. Thus, condition (b) is satisfied with $N = 2$. For points P at which certain scattering directions are prohibited, it may be necessary to take $N > 2$ to satisfy condition (b), but it should still be possible to do so unless the only scattering permitted is delta scattering (i.e. only a single exit direction permitted). The latter is a highly artificial situation, though one which is sometimes approximated in high-energy scattering collisions.

---

† In operator terminology, condition (a) implies that the integral operator $\mathscr{K}$ is bounded in the L_1 norm. Condition (b) implies that $\mathscr{K}_n$ has L_1 norm less than one for $n \geq N$. This condition may be weakened in practice to the condition that $\mathscr{K}_n$ have L_1 norm less than one for some n (Ref. 15).

Even in multiplying media, relation (a) is satisfied since the integral in (3.3.18) is just $c(P')$, the mean number of secondaries per primary at P', which is bounded. Of course, condition (b) no longer follows directly from (a) and one must rely on sufficient absorption or leakage to guarantee (3.3.19).

We shall first establish the main results of this section (Theorems 3.2 and 3.3) making use of the relatively strong assumptions (a) and (b). We then indicate briefly how these assumptions may be weakened to give sharpened results. Our motivation in doing this is to avoid the operator-theoretic arguments which seem necessary in proving the strengthened theorems.

Theorem 3.1 Under the assumptions of conditions (a) and (b),

$$\lim_{m \to \infty} \left\{ \sup_{P_1 \in \Gamma} \int_\Gamma K_m(P, P_1) \, dP \right\} = 0. \tag{3.3.21}$$

Proof. Using (3.3.17),

$$\int_\Gamma K_{2n}(P, P_1) \, dP = \iint_{\Gamma\Gamma} K_n(P, Q) K_n(Q, P_1) \, dQ \, dP.$$

By condition (b), if $n \geq N$, then

$$\iint_{\Gamma\Gamma} K_n(P, Q) K_n(Q, P_1) \, dQ \, dP = \int_\Gamma K_n(Q, P_1) \int_\Gamma K_n(P, Q) \, dP \, dQ$$

$$\leq c \int_\Gamma K_n(Q, P_1) \, dQ \leq c^2,$$

and, more generally,

$$\int_\Gamma K_{ln}(P, P_1) \, dP \leq c^l \qquad \text{for all} \qquad n \geq N.$$

For any integer m, write $m = kN + j$ with $j < N$:

$$\int_\Gamma K_m(P, P_1) \, dP = \iint_{\Gamma\Gamma} K_{kN}(P, Q) K_j(Q, P_1) \, dQ \, dP$$

$$\leq M^j c^k,$$

making use of condition (a). Letting m tend to infinity completes the proof of Theorem 3.1.

Theorem 3.2 Under the assumptions of conditions (a) and (b), the probability model μ based on the analog process (3.3.8)–(3.3.10) is subcritical.

Proof. Using inequality (3.3.13),

$$
\begin{aligned}
\mu(\Lambda_\infty) &\leq \lim_{N \to \infty} \int_\Gamma \cdots \int_\Gamma K(P_N, P_{N-1}) \cdots K(P_2, P_1) S(P_1) \, dP_1 \cdots dP_N \\
&\leq \lim_{N \to \infty} \left[\sup_{P_1 \in \Gamma} \int_\Gamma \cdots \int_\Gamma K(P_N, P_{N-1}) \cdots K(P_2, P_1) \, dP_2 \cdots dP_N \right] \\
&\quad \times \int_\Gamma S(P_1) \, dP_1 \\
&= \lim_{N \to \infty} \left\{ \sup_{P_1 \in \Gamma} \int_\Gamma K_N(P_N, P_1) \, dP_N \right\} = 0 \qquad \text{by Theorem 3.1.}
\end{aligned}
$$

The next result concerns the convergence of the Neumann series in a certain integral sense. For any function $f(P)$, define

$$
\| f(P) \|_1 = \int_\Gamma |f(P)| \, dP \tag{3.3.22}
$$

and

$$
\| \mathscr{K} \|_1 = \sup_{\| f \|_1 \neq 0} \frac{\| \mathscr{K} f \|_1}{\| f \|_1}. \tag{3.3.23}
$$

Theorem 3.3 Conditions (a) and (b) imply the L_1-convergence of the Neumann series to the collision density $\psi(P)$; i.e. if $S_n = S + \mathscr{K}S + \cdots + \mathscr{K}_n S$ denotes the n-th partial sum of the Neumann series, then

$$
\lim_{n \to \infty} \| S_n(P) - \psi(P) \|_1 = 0.
$$

Furthermore, the solution of Eq. (3.3.7) is unique.

Proof. We show that $\| S_m - S_n \|_1 \to 0$ as m and n tend to ∞. Hence, by a well-known result of analysis, the sequence of partial sums is L_1-convergent. Choose $n, m > N$ with $m - n = kN + j \geq 0, j < N$. Then

$$
\begin{aligned}
S_m - S_n &= \mathscr{K}_{n+1}S + \cdots + \mathscr{K}_{n+N}S + \mathscr{K}_{n+N+1}S + \cdots + \mathscr{K}_{n+2N}S \\
&\quad + \cdots + \mathscr{K}_{n+(k-1)N+1}S + \cdots + \mathscr{K}_{n+kN}S + \cdots + \mathscr{K}_{n+kN+j}S \\
&\leq [\mathscr{K}_{n+1} + \cdots + \mathscr{K}_{n+N}]\{S + \mathscr{K}_N S + \cdots + \mathscr{K}_{kN}S\}.
\end{aligned}
$$

Thus,

$$
\| S_m - S_n \|_1 \leq \| \mathscr{K}_{n+1} + \cdots + \mathscr{K}_{n+N} \|_1 \, \| S + \mathscr{K}_N S + \cdots + \mathscr{K}_{kN}S \|_1.
$$

By the triangle inequality for norms ($\| f + g \|_1 \leq \| f \|_1 + \| g \|_1$)

$$
\| S + \mathscr{K}_N S + \cdots + \mathscr{K}_{kN}S \|_1 \leq 1 + c + \cdots + c^k \leq \frac{1}{1 - c}.
$$

Now $\|\mathscr{K}_{n+l}\|_1 \leq \sup_{P_1 \in \Gamma} \int K_{n+l}(P, P_1) \, dP$, so, applying the triangle inequality again and Theorem 3.1, one obtains

$$\|S_m - S_n\|_1 \leq \frac{\epsilon}{1 - c}$$

by choosing n sufficiently large. This proves the L_1-convergence of $\{S_n\}$. Let $\phi(P)$ denote the L_1-limit of the sequence $\{S_n\}$. Then $\phi(P)$ clearly satisfies Eq. (3.3.7). Further, suppose there exists another solution $\psi(P)$. Then

$$\psi(P) - \phi(P) = \int_\Gamma K(P, P')[\psi(P') - \phi(P')] \, dP'$$

$$= \int_\Gamma K_n(P, P')[\psi(P') - \phi(P')] \, dP'$$

for all $n \geq 1$; i.e.

$$\psi(P) - \phi(P) = \mathscr{K}_n[\psi(P) - \phi(P)]$$

for all $n \geq 1$. Choosing $n \geq N$ and taking norms, we have

$$\|\psi - \phi\|_1 \leq \|\mathscr{K}_n\|_1 \|\psi - \phi\|_1 \leq c\|\psi - \phi\|_1,$$

a contradiction unless $\psi = \phi$. This completes the proof of Theorem 3.3.

We have used assumptions (a) and (b) to establish the main results of this section. We now digress briefly to indicate that assumption (b) itself actually implies a condition on the kernel K which allows more direct proofs of the principal results.

This digression requires the introduction of one concept which should be familiar to most readers. The *spectrum* of the integral operator $\mathscr{K}$, $\sigma(\mathscr{K})$, is the set of complex numbers λ such that the operator $\mathscr{K} - \lambda I$ does not have a bounded inverse, I being the identity operator $\big(I[f(P)] = f(P)$ for all functions $f(P)\big)$. The physical process implied by (3.3.7), or the operator $\mathscr{K}$, is called *subcritical* if $\sigma(\mathscr{K})$ is contained in the open unit disk, i.e. if

$$\sigma(\mathscr{K}) \subset \{\lambda \mid |\lambda| < 1\}.$$

The following result may then be proved in a variety of ways (see, e.g., Ref. 15):

Theorem 3.4 Condition (b) implies that $\mathscr{K}$ is subcritical.

Theorems 3.2 and 3.3 actually follow as a result of the assumption that $\mathscr{K}$ be subcritical (Ref. 15), therefore, as a result of assuming condition (b) above. It is the operator-theoretic arguments of Ref. 15 which we have wanted to avoid with the treatment presented above.

We have established a connection between the physical model of Eq. (3.3.7) and the probability model based on the analog random walk process and have shown that the latter probability model is subcritical under natural assumptions on the physical model. We have also shown that these assumptions guarantee the convergence of the Neumann series for the collision density in an integral sense, a fact which will be used often in the discussion to follow.

For probability models based on non-analog random walk processes, an extra assumption will be needed to ensure subcriticality in the sense $\mu(\Lambda_\infty) = 0$. We shall state this condition later in discussing such nonanalog situations.

The probability model given above enables one to construct typical random walks in a manner consistent with the probability function μ. Beginning with an arbitrary random walk process, $\{f_n, p_n\}$, one proceeds as follows (cf. specific discussions in Chapter 2). First, a state P_1 of Γ is chosen from the density function $f_1(P_1)$ according to one of the sampling procedures discussed at length in Chapter 1. Then a decision is made to terminate the walk at P_1 with probability $p_1(P_1)$ or to continue with probability $q_1(P_1) = 1 - p_1(P_1)$. If the decision is made to continue, a state P_2 of Γ is chosen from the conditional density function $f_2(P_2 \mid P_1) = f_2(P_1, P_2)/f_1(P_1)$, where P_1 is the previously determined initial state. If multiplication is allowed, $c(P_1)/q(P_1)$ particles are assigned to the state P_2. In general, if the sequence $P_1, \ldots, P_{n-1}$ of chain points has previously been selected, and if it has been decided not to terminate the walk at P_{n-1} $\left(\text{with probability } q_{n-1}(P_1, \ldots, P_{n-1})\right)$, the next chain point P_n is chosen from the conditional density function

$$f_n(P_n \mid P_1, \ldots, P_{n-1}) = f_n(P_1, \ldots, P_n)/f_{n-1}(P_1, \ldots, P_{n-1}).$$

Then, for each particle of unit weight at P_{n-1}, $c(P_{n-1})/q(P_{n-1})$ particles are assigned to the state P_n. In normal applications $f_n(P_n \mid P_1, \ldots, P_{n-1})$ depends only on P_{n-1} (the Markov case) but our technique is applicable to the more general situation.† This process is repeated until termination of the walk at some point P_k of Γ. The random walk chain which results from this process is the chain

$$C = \{(P_1, \tau), (P_2, \tau), \ldots, (P_{k-1}, \tau), (P_k, \tau'), (P_k, \tau'), (P_k, \tau'), \ldots\}.$$

We see that if $c(P) = q(P)$ for all P, then all particles will carry unit weights. Otherwise, some process (such as Russian roulette and/or splitting, discussed in Section 3.8) may be needed to avoid severe fluctuations in weights and the resulting adverse effects on the variance.

Having briefly described simulation for an arbitrary random walk process $\{f_n, p_n\}$, we now want to discuss somewhat more completely the simulation involved in executing the analog processes (3.3.8)–(3.3.10) and (3.3.8a)–(3.3.10a). Since the first collision density $S(P)$ $\left(\text{Eq. (3.3.8)}\right)$ is often not easy to construct in closed form, one rarely samples $S(P)$ directly. Instead one

---

† Indeed, in certain applications involving Russian roulette (see discussion in Section 3.8) the termination probability may depend on the past history of the particle. We ignore that complication here.

samples $Q(P)$, the true source density, thus selecting birth states. From its birth state, P_0, each sample particle then executes a random walk to its first collision point, P_1. Through such a two-step process $S(P)$ is sampled indirectly.

Similarly, in executing the process governed by Eqs. (3.3.8a)–(3.3.10a), it would generally be difficult to draw sample points directly from the conditional density $L(P_n, P_{n-1})/c'(P_{n-1})$. It would be awkward, for example, to compute the integral

$$c'(P_{n-1}) = \int L(P_n, P_{n-1}) \, dP_n$$

at each collision point. Fortunately, this is not necessary. Again the sampling is usually accomplished indirectly. In fact, the processes characterized by Eqs. (3.3.8)–(3.3.10) and by (3.3.8a)–(3.3.10a) are not essentially different processes. Both processes are usually executed in precisely the same way. One draws a birth point from $Q(P)$, then employs the transport kernel T to find a first collision point. Next the kernel C is used to modify the sample neutron's energy, and T is again used to find a new collision point. Thus, C and T are used, alternately, to generate a random walk: a random walk which we may choose to regard either as a simulation of (2.4.20) or of (3.3.7). In the course of executing such random walks one establishes both ψ and χ. The density of particles about to undergo collisions is ψ, while the density emerging from collisions (or from the source) is χ.

At this point we have discussed at some length the connection between the physical model and the probability model: we have still to define an appropriate estimator ξ whose expected value (3.2.3) gives the required integral (3.2.1). We have given examples of such estimators in Chapter 2, primarily the capture, collision, and track length estimators. Each such estimator is a real-valued function on the space Ω of all random walk chains. The value $\xi(C)$ may be thought of as the weight associated with the particle history C, i.e. the C-estimate of I. This weight is the basic statistical variable of the problem. The basic theoretical problem of Monte Carlo methods is to select estimators ξ whose expectations

$$E[\xi] = \int_\Omega \xi(C) \, d\mu(C) \tag{3.3.24}$$

equal I (i.e. unbiased estimators) and whose variances

$$\sigma^2[\xi] = \int \xi^2(C) \, d\mu(C) - E^2[\xi] \tag{3.3.25}$$

are as small as possible. In practice, however, as mentioned in Section 3.1, theoretical efficiency as measured by low variance must be coupled with computing efficiency, as measured by low cost per history, to achieve best results.

3.4 THE BINOMIAL ESTIMATOR

We have already discussed in Section 3.3 the general notion of an analog random walk process, i.e., one which faithfully simulates a physical model. In the next three sections we shall develop analog estimators more fully, phrasing the definitions within the context of the more general notion of random walk process established in Section 3.3.

As before, the starting point for all of our considerations is the steady-state transport equation. Although we have discussed both the integro-differential form of this equation and two purely integral forms, our purposes in this section are best served by considering the integral form

$$\psi(\mathbf{r}, \mathbf{E}) = \iint_{\Gamma\Gamma} \psi(\mathbf{r}', \mathbf{E}')K(\mathbf{r}, \mathbf{E}; \mathbf{r}', \mathbf{E}') \, d\mathbf{r}' \, d\mathbf{E}' + S(\mathbf{r}, \mathbf{E}) \qquad (3.4.1)$$

for the collision density $\psi(\mathbf{r}, \mathbf{E}) = \Sigma_t(\mathbf{r}, \mathbf{E})\phi(\mathbf{r}, \mathbf{E})$, where Γ is the phase space and $\phi(\mathbf{r}, \mathbf{E})$ is the neutron flux. The analog process for (3.4.1) was defined by Eqs. (3.3.8), (3.3.9), (3.3.10), where we have identified $(\mathbf{r}, \mathbf{E}) = P$.

It is quite commonly the case that the integral

$$I = \int_{\Gamma} g(P)\psi(P) \, dP$$

to be estimated may be written in the form

$$I = \mu(\Lambda), \qquad (3.4.2)$$

where Λ is some subset of the space Ω of all random walks. Then I is interpreted as the probability of Λ as given by the analog probability function μ. For example, if

$$g(P) = \frac{\Sigma_a(P)}{\Sigma_t(P)} \chi_R(P) \qquad (3.4.3)$$

and

$$\chi_R(P) = \begin{cases} 1 & \text{if} \quad P \in R, \\ 0 & \text{if} \quad P \notin R, \end{cases} \qquad (3.4.4)$$

then the integral I is the probability of absorption in the region R of Γ, provided the collision density is normalized to a unit source, as we have assumed previously in Chapter 2. In case the integral I does have the form (3.4.2), it is easy to find an unbiased estimator of I, namely, the function

$$\xi(C) = \begin{cases} 1 & \text{if } C \subset \Lambda, \\ 0 & \text{if } C \notin \Lambda, \end{cases} \qquad (3.4.5)$$

which assigns unity to every random walk in the set Λ and 0 to all others.

Then, by definition,

$$E[\xi] = \int_{\Omega} \xi(C)\,d\mu(C) = 1 \cdot \mu(\Lambda) + 0 \cdot \mu(\Omega - \Lambda) = I,$$

so that ξ is an unbiased estimator of the probability I. The random variable ξ is easily recognized as the binomial estimator associated with the event Λ with probability $I = \mu(\Lambda)$ for "success" and $(1 - I)$ for "failure".

Now, if $C_1, \ldots, C_N$ are N independently chosen random walk chains, the N-tuple $(C_1, \ldots, C_N)$ is a point of the product space Ω^N. On the space Ω^N, N mutually independent random variables ξ_i, $1 \leq i \leq N$, are defined by

$$\xi_i(C_1, \ldots, C_N) = \xi(C_i), \qquad 1 \leq i \leq N, \tag{3.4.6}$$

where ξ is defined by (3.4.5). It follows easily that the average,

$$\bar{\xi}_N = \frac{1}{N} \sum_{i=1}^{N} \xi_i, \tag{3.4.7}$$

is also an unbiased estimator of $I = \mu(\Lambda)$. Since each ξ_i is binomial with variance $I(1 - I)$, ξ_N has variance

$$\sigma^2[\bar{\xi}_N] = \frac{1}{N} I(1 - I) \tag{3.4.8}$$

and, therefore, a relative error which is given by

$$\sigma_r = \frac{\sqrt{\sigma^2[\bar{\xi}_N]}}{I} = \sqrt{\frac{1 - I}{IN}}. \tag{3.4.9}$$

From this expression for the relative error of the estimator $\bar{\xi}_N$ it is easy to see that if I is very small, the sample size N must be very large in order to reduce σ_r to a tolerable value. This matches one's intuition that an accurate estimate of an extremely rare event by use of the binomial estimator requires an inordinately large sample size and, hence, a prohibitive computing cost. This fact motivates the search for alternative unbiased estimators with smaller variances.

In Chapter 2 we discussed the binomial estimator for estimating the absorption rate in a discrete or continuous random walk process. We also discussed collision and track length estimators and made note of the fact that these were frequently more efficient than the binomial estimator. In Section 3.6 of this chapter we shall obtain the collision estimator as a special case of a class of unbiased expected value estimators.

3.5 SYSTEMATIC SOURCE SAMPLING

We shall first describe a device which can always be used to achieve at least a slight decrease in variance. We again assume $I = \mu(\Lambda)$.

Assume that the phase space Γ may be subdivided into a finite number of pair-wise disjoint subregions A_i, and let θ_i be the subset of Ω consisting of all chains C which first collide in A_i. That is,

$$\Gamma = \bigcup_{i=1}^{M} A_i$$

and

$$\theta_i = \{C \in \Omega \mid C = \{\alpha_1, \alpha_2, \ldots\} \text{ and } \alpha_1 = (P_1, \tau) \text{ or } (P_1, \tau') \text{ with } P_1 \in A_i\}.$$

Then

$$\Omega = \bigcup_{i=1}^{M} \theta_i$$

and

$$\sum_{i=1}^{M} \mu(\theta_i) = 1.$$

If we let $p_j = \mu(\theta_j)$, then p_j is the probability of selecting an initial collision point from the region A_j.

Now suppose it has been decided in advance to sample a total of N chains, or some integral multiple kN of chains. In applying the method of systematic source sampling, one selects the initial collisions of the first $p_1 N$ chains from region A_1 (from the conditional density $S(P_1)/p_1 \int_\Gamma S(P) \, dP, P_1 \in A_1$), the initial collisions of the next $p_2 N$ chains from A_2, and so on until all N chains have been selected. This process is repeated k times for the case in which a total of kN chains is desired. For the purposes of the present argument we assume that the numbers $p_i N$ are integers; the nonintegral case may be easily treated by using weighted histories or by using Russian roulette (see Section 3.8).

Now, by Baye's Theorem on conditional probabilities (Ref. 16),

$$I = \mu(\Lambda) = \sum_{i=1}^{M} p_i \mu(\Lambda \mid \theta_i)$$
$$= \sum_{i=1}^{M} p_i I_i, \tag{3.5.1}$$

where $I_i = \mu(\Lambda \mid \theta_i)$ is the conditional probability of Λ given θ_i. One estimates the quantities I_i by using the chains which originate in A_i and then makes use of (3.5.1) and the knowledge of the p_i to obtain an estimate of $\mu(\Lambda)$.

Specifically, we number the N chains $C_1, \ldots, C_N$ so that chains $C_1, \ldots, C_{p_1 N}$ originate in region A_1, etc. Let C_{ij} denote the j-th chain originating in region A_i, $i = 1, \ldots, M$, $j = 1, \ldots, p_i N$. Then define a random variable ξ_i on the space $\Omega^{p_i N}$ by

$$\xi_i(C_{i1}, \ldots, C_{ip_i N}) = \frac{1}{p_i N} \sum_{j=1}^{p_i N} \xi(C_{ij}), \qquad 1 \leq i \leq M, \tag{3.5.2}$$

where

$$\xi(C_{ij}) = \begin{cases} 1 & \text{if } C_{ij} \in \Lambda \\ 0 & \text{if } C_{ij} \notin \Lambda \end{cases} \qquad 1 \leq i \leq M, \qquad 1 \leq j \leq p_i N. \quad (3.5.3)$$

It is easily seen from the definition

$$\mu(\Lambda \mid \theta_i) = \frac{\mu(\Lambda \cap \theta_i)}{\mu(\theta_i)} \quad (3.5.4)$$

that the expected value of ξ_i, when restricted to chains C_{ij} originating in A_i, is $\mu(\Lambda \mid \theta_i) = I_i$. Thus,

$$E[\xi_i] = I_i, \quad (3.5.5)$$

and if we define an estimator ξ'_N on Ω^N by

$$\xi'_N(C_1, \ldots, C_N) = \sum_{i=1}^{M} p_i \xi_i(C_{i1}, \ldots, C_{ip_i N}), \quad (3.5.6)$$

then

$$E[\xi'_N] = \mu(\Lambda) = I,$$

so that ξ'_N is an unbiased estimator of I. The variable ξ'_N is the one used in systematic source sampling.

We now investigate the extent to which the variance of ξ'_N is smaller than the variance (3.4.8) associated with the binomial estimator of I. From Eq. (3.5.6),

$$\sigma^2[\xi'_N] = \sum_{i=1}^{M} p_i^2 \sigma^2[\xi_i] \quad (3.5.7)$$

and, by Eq. (3.5.2),

$$\sigma^2[\xi_i] = \frac{I_i(1 - I_i)}{p_i N} \equiv \frac{\sigma_i^2}{p_i N}. \quad (3.5.8)$$

Then it follows that

$$\sigma^2[\xi'_N] = \sum_{i=1}^{M} \frac{p_i \sigma_i^2}{N}. \quad (3.5.9)$$

Finally, then, the difference $\sigma^2[\bar{\xi}_N] - \sigma^2[\xi'_N]$ is

$$\sigma^2[\bar{\xi}_N] - \sigma^2[\xi'_N] = \frac{1}{N} \sum_{i=1}^{M} p_i(I_i - I)^2. \quad (3.5.10)$$

This quantity, though certainly nonnegative, is rarely a very large fraction of $\sigma^2[\bar{\xi}_N]$. However, since the p_i are ordinarily very easily calculated and the use of the method rarely increases the cost per sample (amount of time necessary to select a chain), it should be used whenever possible.

We have presented the method of systematic source sampling as it would apply to the first collision points of the random walks. Since it is often inconvenient or impractical to precompute the density $S(P)$ of first collisions, the method is often applied to the physical source density in selecting birth

points rather than first collision points systematically. The idea of the
method remains the same and everything said in this section applies subject
to this reinterpretation of first collision point as point of birth. Again, using
the latter point of view, we would study the integral equation for χ rather
than ψ.

3.6 THE USE OF EXPECTED VALUES

In this section we discuss a method which is very widely used, takes many
forms, and has many different names. Most frequently it is called the
"method of expected values" or the "method of statistical estimation." The
idea underlying the method is quite simple. Invariably, in Monte Carlo
work, we are required to compute the mean value of some given estimator.
In expected value methods this mean is evaluated through a combination of
deterministic and random sampling techniques. Consider, for example, a
very simple problem. Suppose we wish to estimate, by Monte Carlo, the
definite integral

$$\theta = \int_0^1 f(x)\, dx = \int_0^1 \int_0^1 g(x, y)\, dx\, dy, \tag{3.6.1}$$

where

$$f(x) = \int_0^1 g(x, y)\, dy. \tag{3.6.2}$$

In method 1 for estimating θ one draws n pairs of random numbers $\rho_1, \rho_2, \ldots,$
ρ_{2n-1}, ρ_{2n} and forms

$$\bar{g}_n = \frac{1}{n} \sum_{i=1}^n g(\rho_{2i-1}, \rho_{2i}). \tag{3.6.3}$$

Then

$$E[\bar{g}_n] = \frac{1}{n} \sum_{i=1}^n \int_0^1 \int_0^1 g(\rho, \rho')\, d\rho\, d\rho'$$
$$= \theta \tag{3.6.4}$$

and

$$\sigma^2[\bar{g}_n] = \frac{1}{n} \int_0^1 \int_0^1 [g - \theta]^2\, dx\, dy$$
$$= \frac{1}{n} \int_0^1 \int_0^1 g^2(x, y)\, dx\, dy - \frac{\theta^2}{n}. \tag{3.6.5}$$

In method 2, an expected value method, one draws n random numbers
$\rho_1, \ldots, \rho_n$ and forms

$$\bar{f}_n = \frac{1}{n} \sum_{i=1}^n f(\rho_i). \tag{3.6.6}$$

As before, $E[\bar{f}_n] = \theta$, but

$$\sigma^2[\bar{f}_n] = \frac{1}{n} \int_0^1 f^2(x)\, dx - \frac{\theta^2}{n} \tag{3.6.7}$$

and

$$\sigma^2[\tilde{g}_n] - \sigma^2[\tilde{f}_n] = \frac{1}{n}\left[\int_0^1\int_0^1 g^2(x, y)\, dx\, dy - \int_0^1 f^2(x)\, dx\right]$$

$$= \frac{1}{n}\left\{\int_0^1\int_0^1 g^2(x, y)\, dx\, dy - \int_0^1\left[\int_0^1 g(x, y)\, dy\right]^2 dx\right\}$$

$$= \frac{1}{n}\int_0^1 dx\left\{\int_1^1 g^2(x, y)\, dy - \left[\int_0^1 g(x, y)\, dy\right]^2\right\}$$

$$= \frac{1}{n}\int_0^1 dx\int_0^1 dy[g - f]^2 \geq 0.$$

This inequality shows that method 2 is superior to method 1 from the point of view of variance, as one would expect since the variability of $g(x, y)$ with respect to y has been removed.

Note that, in method 2, values of x are sampled statistically, but for each given value of x the estimator is averaged deterministically over all possible y-values. If the selection of an (x, y)-pair is regarded as an "event", then, having fixed x, we have taken account of all events which *might have* occurred. Deterministic averaging over "potential events" is the characteristic feature of expected value methods.

It is important to realize that, in our past work, we have already discussed several different expected value estimators. All the estimators of Section 2.7 are expected value estimators. Suppose we wish to compute the absorption rate in region A. Imagine that a sample particle has just made a collision in A, or has just entered A. Subsequently, on the way to its next collision it leaves, in region A, a track of length d. The quantity $\Sigma_a d$ is, then, the resulting contribution to the track length estimator. But we may choose to record, instead, the quantity (see p. 80)

$$E_{abc} = \frac{\Sigma_a}{\Sigma_t}(1 - e^{-\Sigma_t D}),$$

i.e. the *expected* contribution to the track length estimator. We will, then, have averaged the track length estimator, deterministically, over a set of *possible* flight paths. In this sense E_{abc} is an expected value estimator. The reader will recall that the estimator E_{abc} does not necessarily have a smaller variance per history than the original track length estimator. Thus, we do not necessarily reduce the variance per history through the use of expected values, though, generally, the variance per history is reduced and this is, of course, the hope when the technique is applied.

Wasow's estimator (see Eq. 2.5.8) is a somewhat different expected value estimator. Suppose, again, that we wish to compute the absorption rate in region A. When we use a binomial estimator, we record the true number of absorptions which occur at each collision point. On the other hand, in using

Wasow's estimator we record the expected number of absorptions at each collision point. Thus, again, we take into account not only the events which actually occur during the course of each history but also various events which might have occurred. Though Wasow's estimator was introduced in Chapter 2, we have not yet shown that it is unbiased, and we shall do this next.

We begin, as usual, with the integral equation for the collision density

$$\psi(P) = \int_\Gamma K(P, P')\psi(P')\, dP' + S(P) \tag{3.6.8}$$

subject to the restrictions $S \geq 0$, $K \geq 0$, $\int_\Gamma S(P)\, dP = 1$, and conditions (a) and (b) (Eqs. 3.3.18, 3.3.19) of Section 3.3. It follows from Theorem 3.4 that these conditions imply the convergence of the series of integrals of the Neumann series:

$$\int \psi(P)\, dP = \int S(P)\, dP + \iint K(P, P_1)S(P_1)\, dP_1\, dP + \cdots.$$

Let $g(P)$ be any bounded function so that the series

$$\int g(P)S(P)\, dP + \iint g(P)K(P, P_1)S(P_1)\, dP_1\, dP + \cdots$$

also converges to $\int g(P)\psi(P)\, dP = I$. We are now equipped to prove the following theorem.† The estimator of the theorem generalizes Wasow's estimator to continuous, multiplying processes.

Theorem 3.5 For the analog process defined by S and K (Eqs. 3.3.8–3.3.10) the random variable

$$\xi(P_1, \ldots, P_k) = \sum_{i=1}^{k} \left[g(P_i) \prod_{j=1}^{i-1} \frac{c(P_j)}{q(P_j)} \right] \tag{3.6.9}$$

is an unbiased estimator of I. In the theorem the points $P_1, \ldots, P_k$ represent the collision points of the random walk, which terminates at P_k, and the ratio $c(P_j)/q(P_j)$ may be regarded as a multiplicative weight acquired upon collision at P_j which accounts for the possibility of multiplication (cf. Eq. 3.3.11).

Proof. Let μ denote the analog probability function. Then

$$\begin{aligned}
E[\xi] &= \int_\Omega \xi(C)\, d\mu(C) = \sum_{k=1}^{\infty} \int_{\Lambda_k} \xi(C)\, d\mu(C) \\
&= \sum_{k=1}^{\infty} \mu(\Lambda_k) \int_{\Lambda_k} \xi(C)\, d\mu(C)/\mu(\Lambda_k) \\
&= \sum_{k=1}^{\infty} \mu(\Lambda_k) E[\xi \mid k(C) = k],
\end{aligned}$$

---

† See also Ref. 15.

where $E[\xi \mid k(C) = k]$ is the conditional expected value of ξ restricted to chains of length k.

By the definition of the analog probability function μ on chains of length k,

$$E[\xi \mid k(C) = k] = \frac{1}{\mu(\Lambda_k)} \int_\Gamma \cdots \int_\Gamma \xi(P_1, \ldots, P_k) \prod_{j=2}^{k} K(P_j, P_{j-1})$$

$$\times S(P_1) \prod_{j=1}^{k-1} \frac{q(P_j)}{c(P_j)} p(P_k) \, dP_1 \cdots dP_k.$$

Then

$$E[\xi] = \sum_{k=1}^{\infty} \sum_{i=1}^{k} \int_\Gamma \cdots \int_\Gamma \left[g(P_i) \prod_{j=1}^{i-1} [c(P_j)/q(P_j)] \right] \prod_{j=2}^{k} K(P_j, P_{j-1})$$

$$\times S(P_1) \prod_{j=1}^{k-1} [q(P_j)/c(P_j)]p(P_k) \, dP_1 \cdots P_k.$$

Now the double sum $\sum_{k=1}^{\infty} \sum_{i=1}^{k}$ may be permuted to $\sum_{i=1}^{\infty} \sum_{k=i}^{\infty}$, which results in

$$E[\xi] = \sum_{i=1}^{\infty} \sum_{k=i}^{\infty} \int_\Gamma \cdots \int_\Gamma \left[g(P_i) \prod_{j=1}^{i-1} c(P_j)/q(P_j) \right] \prod_{j=2}^{k} K(P_j, P_{j-1})$$

$$\times S(P_1) \prod_{j=1}^{k-1} [q(P_j)/c(P_j)]p(P_k) \, dP_1 \cdots dP_k$$

$$= \sum_{i=1}^{\infty} \int_\Gamma \cdots \int_\Gamma \left[g(P_i) \prod_{j=1}^{i-1} c(P_j)/q(P_j) \right] \prod_{j=2}^{i} K(P_j, P_{j-i1})$$

$$\times S(P_1) \prod_{j=1}^{i-1} [q(P_j)/c(P_j)] \, dP_1 \cdots P_i$$

$$\times \left\{ p(P_i) + \int_\Gamma K(P_{i+1}, P_i)[q(P_i)/c(P_i)]p(P_{i+1}) \, dP_{i+1} + \cdots \right\}.$$

Since $p(P) + q(P) = 1$, the infinite series in braces reduces to

$$p(P_i) + q(P_i) - \lim_{k \to \infty} \int_\Gamma \cdots \int_\Gamma K(P_k, P_{k-1}) \cdots K(P_{i+1}, P_i)$$

$$\times \prod_{j=1}^{k} [q(P_j)/c(P_j)] \, dP_{i+1} \cdots dP_k$$

$$= 1 - \lim_{k \to \infty} \int_\Gamma \cdots \int_\Gamma K(P_k, P_{k-1}) \cdots K(P_{i+1}, P_i)$$

$$\times \prod_{j=i}^{k} [q(P_j)/c(P_j)] \, dP_{i+1} \cdots dP_k$$

$$= 1$$

by Theorem 3.1 and the fact that $\sup_{P \in \Gamma} q(P)/c(P) \leq 1$. Then

$$E[\xi] = \sum_{i=1}^{\infty} \int_{\Gamma} \cdots \int_{\Gamma} g(P_i) \prod_{j=2}^{i} K(P_j, P_{j-1}) S(P_1) \, dP_1 \cdots dP_i$$
$$= I$$

by Theorem 3.3 and by the boundedness of g.

The previous theorem on statistical estimation is an interesting one. Although we did not make use of the fact in the proof, it may be shown that, for fixed P_1, the random variable

$$\xi(P_1, P_2, \ldots, P_k) = \sum_{i=1}^{k} \left[g(P_i) \prod_{j=1}^{i-1} c(P_j)/q(P_j) \right] \tag{3.6.10}$$

is an unbiased estimator of the adjoint collision density $\psi^*(P)$ evaluated at the point P_1; i.e. the function $\psi^*(P)$ which satisfies the adjoint integral equation

$$\psi^*(P) = \int K(P', P) \psi^*(P') \, dP' + g(P). \tag{3.6.11}$$

This fact would yield an alternative proof of the theorem in as much as the expected value of ξ with respect to P_1 would be

$$E[\xi] = \int \psi^*(P_1) S(P_1) \, dP_1 = \int g(P_1) \psi(P_1) \, dP_1 = I \tag{3.6.12}$$

by making use of reciprocity. This fact also makes plausible the interpretation of the estimator ξ as a generalized expected value estimator. Indeed, by (3.6.12), $\psi^*(P_1)$ may be regarded as the expected contribution to I from a particle whose first collision is at P_1. The adjoint Eq. (3.6.11) for $\psi^*(P)$ admits the following interpretation: the total expected contribution to I from P is comprised of an expected direct contribution, $g(P)$ (no further collisions), plus an indirect contribution, the integral term, which represents the contribution to I from P from particles which have collided at P'. In this interpretation, which is very closely related to the interpretation of $\psi^*(P)$ as an importance function for the estimation of I, $g(P)$ is the expected direct contribution from P and it is this expected direct contribution appropriately weighted and summed over all collision points, which defines the expected value estimator ξ.

If we apply Theorem 3.5 to a non-multiplying random walk process $\left(c(P) = q(P) = \Sigma_s(P)/\Sigma_t(P)\right)$ and to the choice

$$g(P) = \frac{\Sigma(P)}{\Sigma_t(P)},$$

where $\Sigma(P)$ is an arbitrary macroscopic cross section, we obtain the collision estimator (Eqs. 2.4.29, 2.6.3) discussed in Chapter 2. Theorem 3.5 thus provides a proof that the collision estimator is unbiased.

Theorem 3.5 also gives another important result obtained by setting $g(P) = 1$. Then the random variable

$$\xi(P_1, \ldots, P_k) = \sum_{i=1}^{k} \prod_{j=1}^{i-1} \frac{c(P_j)}{q(P_j)}$$

is an unbiased estimator of $\int_{\Gamma} \psi(P) \, dP$, the expected weight of collisions made by all particles. (Note that if $c(P) = q(P)$, then $\xi = k$, the number of collisions, as one expects.) The fact that this expected weight of collisions is finite follows from Ref. 15 under the weaker assumption that $\mathscr{K}$ be subcritical. This result, which is stronger than $\mu(\Lambda_\infty) = 0$ $\big(\text{since } c(P) \geq q(P)\big)$, is important from a computational point of view and establishes a closer tie between the physical and probability models.

As Theorem 3.5 has been presented, it deals with a random variable which records at the discrete collision points of the neutron history, excluding its birth point. Therefore, in applying Theorem 3.5, no contributions are counted unless and until the particle has collided at a given point. However, the term "statistical estimation" has also been used to suggest methods in which a particle is made to contribute to regions along its direction of flight, *prior* to any actual collision at points of such regions. This point of view is especially useful in, for example, shielding problems where deep penetration is an unlikely event. We shall see that such estimators may be derived directly by considering the integral Eq. (2.4.20) for the function χ rather than the Eq. (3.6.8) for the collision density ψ.

Indeed, if one uses the relationship (2.4.19) between ψ and χ,

$$\psi(\mathbf{r}, \mathbf{E}) = \int \chi(\mathbf{r}', \mathbf{E}) T(\mathbf{r}', \mathbf{r}; \mathbf{E}) \, d\mathbf{r}', \tag{3.6.13}$$

then one sees that the integral I may also be written

$$I = \int_{\Gamma} g(P)\psi(P) \, dP = \iiint g(\mathbf{r}, \mathbf{E})\chi(\mathbf{r}', \mathbf{E})T(\mathbf{r}', \mathbf{r}; \mathbf{E}) \, d\mathbf{r}' \, d\mathbf{r} \, d\mathbf{E},$$

so that

$$I = \iint f(\mathbf{r}', \mathbf{E})\chi(\mathbf{r}', \mathbf{E}) \, d\mathbf{r}' \, d\mathbf{E}, \tag{3.6.14}$$

where

$$f(P) = f(\mathbf{r}, \mathbf{E}) = \int g(\mathbf{r}', \mathbf{E})T(\mathbf{r}, \mathbf{r}'; \mathbf{E}) \, d\mathbf{r}'. \tag{3.6.15}$$

The function f gives the expected contribution to I from a neutron which has just undergone collision at $(\mathbf{r}, \mathbf{E})$ provided no collisions intervene between $\mathbf{r}$ and $\mathbf{r}'$. Then the analog of Theorem 3.5 yields an estimator

$$\eta(P_0, \ldots, P_k) = \sum_{i=0}^{k} \left[f(P_i) \prod_{j=0}^{i-1} \frac{c(P_j)}{q(P_j)} \right] \tag{3.6.16}$$

of I, where we have written P_0 instead of P_1 to emphasize that P_0 is selected from the physical source density Q, not the density of first collisions, S.

We note that if

$$g(P) = \frac{\Sigma(P)}{\Sigma_t(P)} \chi_R(P),$$

where $\chi_R(P)$ is the characteristic function of a region R of constant cross section, then I is the Σ-reaction rate in R. In this case,

$$f(P) = \int \frac{\Sigma(\mathbf{r}', E)}{\Sigma_t(\mathbf{r}', E)} \chi_R(\mathbf{r}', E) \Sigma_t(\mathbf{r}', E) \exp\left[-\int_0^{\boldsymbol{\omega} \cdot (\mathbf{r}' - \mathbf{r})} \Sigma_t(\mathbf{r} + s\boldsymbol{\omega}, E)\, ds\right] d\mathbf{r}'$$

$$= \frac{\Sigma(E)}{\Sigma_t(E)} \exp\left[-\int_0^{D_{\min}} \Sigma_t(\mathbf{r} + s\boldsymbol{\omega}, E)\, ds\right]$$

$$\times \left[1 - \exp\left(-\Sigma_t(E)(D_{\max} - D_{\min})\right)\right], \quad (3.6.17)$$

where $D_{\min}$ is the minimum distance from r to R along $\boldsymbol{\omega}$ and $D_{\max}$ is the largest possible such distance, provided the ray $(\mathbf{r} + s\boldsymbol{\omega})$ intersects R only once. The exponential integral in (3.6.17) is usually obtained as a sum of exponentials, one exponential arising from each distinct region of constant cross section lying between $\mathbf{r}$ and the region R. If more than one such intersection takes place, a contribution like (3.6.17) is obtained from each potential crossing. This points out the essential difficulty of using such an estimator with cell boundary conditions, in which case an infinite number of crossings may arise from each collision.

If we rewrite (3.6.17) as

$$f(P) = \left\{\exp\left[-\int_0^{D_{\min}} \Sigma_t(\mathbf{r} + s\boldsymbol{\omega}, E)\, ds\right]\right\}$$

$$\times \left\{\frac{\Sigma(E)}{\Sigma_t(E)}\left[1 - \exp\left(-\Sigma_t(E)(D_{\max} - D_{\min})\right)\right]\right\}, \quad (3.6.18)$$

then we recognize this expression as the product of the probability of reaching R times the estimator E_{abc} of Section 2.7. In this form we see that the second factor is no more than a modified track length estimator of the Σ-reaction rate in R, and may be replaced by any other estimator of the same quantity. It is clear that a function such as (3.6.18) may not be used to estimate reaction rate at a point, since, in this limiting case, it will give a contribution with probability zero. Nevertheless, it may be an extremely useful estimator when applied to small regions.

We wish to point out that this process of taking expected values over preceding events may be carried out, in principle, any number of times. At each successive stage, of course, the calculation becomes progressively more difficult. It is instructive, however, to try to carry out one more stage of this process. To do so, we make use of the equation

$$\chi(\mathbf{r}, E) - Q(\mathbf{r}, E) = \int \psi(\mathbf{r}, E')C(E', E; \mathbf{r}) \, dE' \tag{3.6.19}$$

so that

$$I = \iint f(\mathbf{r}, E)\chi(\mathbf{r}, E) \, d\mathbf{r} \, dE$$

$$= \iint f(\mathbf{r}, E)Q(\mathbf{r}, E) \, d\mathbf{r} \, dE + \iiint f(\mathbf{r}, E)C(E', E; \mathbf{r})\psi(\mathbf{r}, E') \, dE' \, dE \, d\mathbf{r}.$$

If we then define

$$h(\mathbf{r}, E') = \int f(\mathbf{r}, E)C(E', E; \mathbf{r}) \, dE$$

$$= \iint g(\mathbf{r}', E)T(\mathbf{r}, \mathbf{r}'; E)C(E', E; \mathbf{r}) \, d\mathbf{r}' \, dE, \tag{3.6.20}$$

we see that the function h may be used in place of g in Theorem 3.5. The estimator (3.6.10) with g replaced by h, when added to the uncollided contribution $\iint f(\mathbf{r}, E)Q(\mathbf{r}, E) \, d\mathbf{r} \, dE$, would again give an unbiased estimator of I. Notice that the effect of using h is to examine at $\mathbf{r}$ all possible collisions which result in final energy vectors E, then to transfer particles from $\mathbf{r}$ to $\mathbf{r}'$, and finally to make contributions at $(\mathbf{r}', E)$.

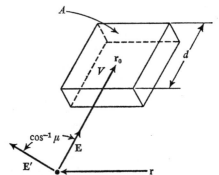

Fig. 3.1

Suppose we attempt to make use of (3.6.20) to estimate, say, the flux at a point $\mathbf{r}_0$. Let V denote a small volume about $\mathbf{r}_0$, as depicted in Fig. 3.1, with cross sectional area A, and length d along the direction of $\mathbf{r}_0$. We take

$$g(\mathbf{r}', E) = \frac{1}{\Sigma_t(\mathbf{r}', E)} \chi_V(\mathbf{r}')/V \tag{3.6.21}$$

so that I is the flux integral per unit volume over V. It follows from (3.6.20) that

$$h(\mathbf{r}, E') = \iint \frac{\chi_V(\mathbf{r}')}{\Sigma_t(\mathbf{r}', E)V} \Sigma_t(\mathbf{r}', E)$$

$$\times \exp\left[-\int_0^{\boldsymbol{\omega}\cdot(\mathbf{r}'-\mathbf{r})} \Sigma_t(\mathbf{r} + s\boldsymbol{\omega})\, ds\right] C(E', E; \mathbf{r})\, d\mathbf{r}'\, d\mathbf{E}.$$

Note that the total cross section $\Sigma_t(\mathbf{r} + s\boldsymbol{\omega})$ is to be evaluated at the energy E. Now let

$$C(E', E; \mathbf{r}) = c(E', \mathbf{r})P(E' \to E; \mathbf{r} \mid \boldsymbol{\omega}' \to \boldsymbol{\omega})f_{E',\mathbf{r}}(\boldsymbol{\omega}' \to \boldsymbol{\omega}).$$

Here $c(E', \mathbf{r})$ is the number of secondaries per collision,

$$c(E', \mathbf{r}) = \int C(E', E; \mathbf{r})\, dE,$$

while $f_{E',\mathbf{r}}(\boldsymbol{\omega}' \to \boldsymbol{\omega})$ is the probability that a neutron colliding at $\mathbf{r}$ with energy E', will be scattered from $\boldsymbol{\omega}'$ into a unit solid angle about $\boldsymbol{\omega}$. Finally, $P(E' \to E; r \mid \boldsymbol{\omega}' \to \boldsymbol{\omega})$ is the conditional probability that, in a collision which takes $\boldsymbol{\omega}'$ into $\boldsymbol{\omega}$, the neutron's energy will change from E' to E. Now it is clear that V subtends, at $\mathbf{r}$, the solid angle A/t^2, where $t = |\mathbf{r} - \mathbf{r}_0|$. Therefore

$$h(\mathbf{r}, E') = \left\{\int dE \exp\left[-\int_0^{\boldsymbol{\omega}_0\cdot(\mathbf{r}_0-\mathbf{r})} \Sigma_t(\mathbf{r} + s\boldsymbol{\omega})\, ds\right] P(E' \to E; \mathbf{r} \mid \boldsymbol{\omega}' \to \boldsymbol{\omega})\right\}$$

$$\times f_{E',\mathbf{r}}(\boldsymbol{\omega}' \to \boldsymbol{\omega}_0)c(E', \mathbf{r})A d/t^2 V,$$

$$= \left\{\int dE \exp\left[-\int_0^{\boldsymbol{\omega}_0\cdot(\mathbf{r}_0-\mathbf{r})} \Sigma_t(\mathbf{r} + s\boldsymbol{\omega})P(E' \to E; \mathbf{r} \mid \boldsymbol{\omega}' \to \boldsymbol{\omega})\right\}\right.$$

$$\times f_{E;\mathbf{r}}(\boldsymbol{\omega}' \to \boldsymbol{\omega}_0)c(E', \mathbf{r})/t^2. \tag{3.6.22}$$

In (3.6.22) the vector $\boldsymbol{\omega}_0$ is, of course, a unit vector directed from $\mathbf{r}$ to $\mathbf{r}_0$.

The function $h(\mathbf{r}, E')$ is represented, in (3.6.22), as an integral over E. It may appear that this integral would be difficult to evaluate in the course of a Monte Carlo calculation. Often, however, this is not the case. In an *elastic* collision with any stationary nucleus the final neutron energy, E_0, is completely determined† by E', $\boldsymbol{\omega}$, and $\boldsymbol{\omega}'$; i.e.

$$E_0 = E'\{\sqrt{A^2 - 1 + \mu^2} + \mu\}^2/(A + 1)^2.$$

Here $\mu = \boldsymbol{\omega}' \cdot \boldsymbol{\omega}$, and A is the ratio of the mass of the nucleus to the mass of the neutron. Thus, if only one isotope were present at $\mathbf{r}$, and if elastic scattering and absorption were the only physical processes operative at

---

† See S. Glasstone and M. C. Edlund, *The Elements of Nuclear Reactor Theory*, D. Van Nostrand Co. (1952).

incoming energy E', then we would have

$$P(E' \to E; \mathbf{r} \mid \boldsymbol{\omega}' \to \boldsymbol{\omega}) = \delta(E' - E_0),$$

so that

$$h(\mathbf{r}, E') = c(E', \mathbf{r}) f_{E', \mathbf{r}}(\boldsymbol{\omega}' \to \boldsymbol{\omega}_0) \exp\left[-\int_0^{\boldsymbol{\omega}_0 \cdot (\mathbf{r}_0 - \mathbf{r})} \Sigma_t(\mathbf{r} + s\boldsymbol{\omega}_0) \, ds \right] \bigg/ t^2.$$

$$(3.6.23)$$

It is not difficult to generalize Eq. (3.6.23) to situations where several elastic scatterers are present at $\mathbf{r}$, and even to cases where fission at $\mathbf{r}$ is allowed.

As Kalos (Ref. 17) has pointed out, an estimator based on (3.6.23) has infinite variance because of the $1/t^2$ singularity near $t = 0$; i.e. collisions near the point $\mathbf{r}_0$ will result not only in very large contributions to the flux but also in large fluctuations about the average. Kalos shows that an unbiased and convergent procedure results despite the infinite variance. He also shows how finite variance estimators may be derived by modifying (3.6.23) through the introduction of an intermediate collision point and by proper averaging over such collision points. Such estimators, however, have practical disadvantages when applied to problems with great geometric detail.

3.7 IMPORTANCE SAMPLING

We turn in this section to a class of non-analog methods for obtaining estimators with reduced variances.

Stated briefly, the notion of importance sampling involves a distortion of the actual physical transition probabilities (i.e. a distortion of the analog random walk process as defined in Section 3.3) with the net result that events of interest in the calculation will occur more frequently than in the analog process. This distortion is then compensated by a corresponding alteration of the estimating random variable so as to remove any bias from the estimates of the quantities of interest.

Through the use of importance sampling one can (in principle) not only reduce variance but actually eliminate variance. Invoking the theory of importance sampling, one can develop zero-variance estimation techniques for solving problems of many types. Zero-variance methods are Monte Carlo estimation methods, but they are specifically designed to give *exact* estimates completely free of statistical fluctuations. It is possible, for example, to devise zero-variance importance sampling methods for the computation of neutron reaction rates. In a zero-variance computation of some specific reaction rate all histories yield identical, and identically correct, estimates of this single quantity.

Unfortunately, the zero-variance schemes cannot be used, directly, to solve practical problems. To formulate a zero-variance estimation scheme, we must already know the value of the quantity which we propose to estimate; we cannot estimate a reaction rate without error unless we already know the reaction rate. Of course, if we have such good information we can construct many different zero-variance schemes which are all perfectly valid. Importance sampling leads us, then, to one of many alternatives, and it is an alternative which, *a priori*, does not seem particularly useful. The importance sampling zero-variance scheme involves not only knowledge of the unknown reaction rate but also knowledge of the neutron flux (or its adjoint) everywhere in the problem configuration.

Nevertheless, this scheme is not merely a mathematical curiosity. Zero-variance importance sampling is an optimum Monte Carlo strategy, an optimum which cannot be attained in practice, but one which can be approached. If we know the flux (or its adjoint) *approximately*, we can formulate a Monte Carlo strategy which is nearly optimum. Thus, we have, in importance sampling, a way to factor prior knowledge into Monte Carlo, a way to make use of approximate solutions of the transport equation. The capacity to use estimated or guessed solutions is lacking in other Monte Carlo methods, though it is a common feature of deterministic methods. For example, in iterative methods for solving differential equations, or matrix equations, one can usually start the chain of iterations from an estimated or guessed solution. If the guess is good, the iterative process should converge very quickly. Their ability to utilize prior information often gives iterative methods a great advantage over noniterative methods and it seems desirable to introduce this same capability into Monte Carlo.

The idea of applying importance sampling to transport problems and the theory of zero-variance importance sampling seems to originate with Goertzel (Ref. 18). More recently, Kahn has given expositions of the method in Refs. 19 and 20, and Goertzel and Kalos in Refs. 21 and 22. Albert has discussed in Ref. 23 a method of estimating Neumann series and terms of Neumann series at a specific point, a method which bears a strong resemblance to importance sampling. The point of view presented in this section unifies the results of Kahn and Albert.

As usual, we consider the problem of estimating the integral

$$I = \int g(P)\psi(P)\, dP, \qquad (3.7.1)$$

where $\psi(P)$ is the collision density.† We have already seen that, in an analog calculation, the estimation of the integral I is accomplished by estimation of

---

† As we have seen already (pp. 106ff), it is sometimes advantageous to work with integrals of χ rather than ψ. For the moment, since g usually takes a particularly simple form in (3.7.1), we shall work with the collision density.

the expected value

$$E[\xi] = \int_\Omega \xi(C)\, d\mu(C), \qquad (3.7.2)$$

where ξ is an unbiased estimator of I and μ is the analog probability function defined on Ω. Formally, a change of variable may be made and the integral (3.7.2) rewritten in the form

$$E[\xi] = \int_\Omega \left[\xi \frac{d\mu}{d\hat{\mu}}\right] d\hat{\mu}, \qquad (3.7.3)$$

where $d\mu/d\hat{\mu}$ is a function which represents the generalized (Radon-Nikodym) derivative† of μ with respect to $\hat{\mu}$. In practice, such integrals will turn out to be series of ordinary Riemann integrals. Then, if

$$\hat{\xi}(C) \equiv \xi(C) \frac{d\mu}{d\hat{\mu}}(C), \qquad (3.7.4)$$

one has

$$E[\hat{\xi}] = \int_\Omega \hat{\xi}(C)\, d\hat{\mu}(C). \qquad (3.7.5)$$

This shows that the random variable $\hat{\xi}$ is an unbiased estimator of I with respect to the new probability function $\hat{\mu}$ on Ω. This is precisely the basis for importance sampling. The new probability function $\hat{\mu}$ on Ω is defined in terms of a random walk process $\{\hat{f}_n, \hat{p}_n\}$ which is selected to emphasize those events which are important to the estimation of I. This is accomplished by a method analogous to that used in defining the analog probability function. The new estimator $\hat{\xi}$ defined by (3.7.4) is then an unbiased estimator of I, provided ξ is unbiased with respect to the analog probability function μ.

The question now naturally arises: How is the random walk process $\{\hat{f}_n, \hat{p}_n\}$ selected so as to minimize the variance of the random variable $\hat{\xi}$? We have implied that the random walk process $\{\hat{f}_n, \hat{p}_n\}$ may be arbitrary but this is not quite correct, since some restriction is necessary to guarantee that the generalized derivative $d\mu/d\hat{\mu}$ (hence, $\hat{\xi}$) should exist. We also need to guarantee that the probability model based on such a random walk process be subcritical in the sense of Eq. (3.3.6). In most cases the new random walk process is defined by formulas such as (3.3.8)–(3.3.10), except that a new source $\hat{S}(P)$ and a new kernel $\hat{K}(P, P')$ are used. In this case the source $\hat{S}$ and kernel $\hat{K}$ may be regarded as defining a new integral transport equation for a transformed collision density $\hat{\psi}(P)$:

$$\hat{\psi}(P) = \int_\Gamma \hat{K}(P, P')\hat{\psi}(P')\, dP' + \hat{S}(P). \qquad (3.7.6)$$

---

† Those readers familiar with this concept will realize that a very compact notation and description result from its use. In any event, the discussion in the remainder of this section will be explicit and will not require any familiarity with the Radon-Nikodym derivative.

In Eq. (3.7.6) the source $\hat{S}$ is used to construct a distribution of initial collision points and the kernel $\hat{K}$ is used to move particles from state to state. However, the particles may be thought of as carrying weights in order to adjust the expected weight undergoing collision to the analog value. Thus, a particle which suffers an initial collision at P_1 must be assigned a weight $S(P_1)/\hat{S}(P_1)$, and a particle which moves from P_i to P_{i+1} has its weight multiplied by the factor

$$\frac{K(P_{i+1}, P_i)}{\hat{K}(P_{i+1}, P_i)} \frac{\hat{c}(P_i)}{\hat{q}_i(P_1, \ldots, P_i)} .$$

Finally, when a history terminates at P_k, its weight is multiplied by

$$\frac{p(P_k)}{\hat{p}_k(P_1, \ldots, P_k)} .$$

Then, if $W(P)$ denotes the expected weight density of particles undergoing collision at P, $W(P)$ may be shown to satisfy the same equation as $\psi(P)$, and, hence, by uniqueness, $W(P) = \psi(P)$. The foregoing gives a convenient interpretation of how weights may be used to readjust the collision density to its analog value. We shall usually regard the weights as incorporated in the estimating random variable rather than as an intrinsic part of the underlying physical process.

We shall require that the source $\hat{S}$ and kernel $\hat{K}$ satisfy the usual conditions $\hat{S}, \hat{K} \geq 0$, $\int_\Gamma \hat{S}(P) \, dP = 1$, and the conditions (a) and (b), Eqs. (3.3.18) and (3.3.19) of Section 3.3. (Again, as was pointed out in Section 3.3, these conditions may be replaced by the weaker condition that the integral operator $\hat{\mathscr{K}}$ be subcritical.)

Conditions needed to establish that the generalized derivative $d\mu/d\hat{\mu}$ exist are

$$\begin{cases} \hat{S}(P) = 0 \Rightarrow S(P) = 0, \\[4pt] \dfrac{\hat{K}(P, P')}{\hat{c}(P')} = 0 \Rightarrow \dfrac{K(P, P')}{c(P)} = 0, \\[4pt] \hat{p}_j(P_1, \ldots, P_{j-1}, P) = 0 \Rightarrow p(P) = 0, \\[4pt] \hat{q}_j(P_1, \ldots, P_{j-1}, P) = 0 \Rightarrow q(P) = 0. \end{cases} \qquad (3.7.7)$$

Furthermore, for $C = (P_1, \ldots, P_k, P_k, \ldots) \in \Lambda_k$, the function $X(P_1, \ldots, P_k)$

$$= \prod_{j=2}^{k} \frac{K(P_j, P_{j-1})}{\hat{K}(P_j, P_{j-1})} \frac{q(P_{j-1})}{\hat{q}_{j-1}(P_1, \ldots, P_{j-1})} \frac{\hat{c}(P_{j-1})}{c(P_{j-1})} \frac{S(P_1)}{\hat{S}(P_1)} \frac{p(P_k)}{\hat{p}_k(P_1, \ldots, P_k)} \qquad (3.7.8)$$

must be bounded except possibly for sets of chains of probability zero in the $\hat{\mu}$

process. In the above $\hat{c}(P) = \int_\Gamma \hat{K}(Q, P)\, dQ$ is the mean number of second-aries per primary on collision at P for the kernel $\hat{K}$. The existence of $d\mu/d\hat{\mu}$ under these conditions is proved in Ref. 15; because of its rather technical nature we shall not emphasize it here.

A further condition is also needed for such non-analog processes, namely, condition (c) below.

c) There exists an integer M such that, for $n \geq M$,

$$\prod_{i=1}^{n} \frac{\hat{q}_i(P_1, \ldots, P_i)}{\hat{c}(P_i)} \leq 1 \qquad \text{for all} \qquad P_1, \ldots, P_n. \qquad (3.7.9)$$

Making use of Eq. (3.3.13), one sees that this condition guarantees that $\hat{\mu}(\Lambda_\infty) = 0$ for the probability function $\hat{\mu}$ defined by the new random walk process.

All of the following theorems assume that $\hat{\mathcal{K}}$ is subcritical, that $d\mu/d\hat{\mu}$ exists, and that condition (c) (Eq. 3.7.9) holds. The next theorem we prove generalizes the estimator of Eq. (2.5.5) to multiplying media.

Theorem 3.6 The random variable $\hat{I}(C)$ defined on chains $C = \{P_1, \ldots, P_n\}$ of length n by

$$\hat{I}(C) = \frac{g(P_n)K(P_n, P_{n-1}) \cdots K(P_2, P_1)S(P_1)}{\hat{K}(P_n, P_{n-1}) \cdots \hat{K}(P_2, P_1)\hat{S}(P_1)\hat{p}(P_n)} \prod_{j=1}^{n-1} \frac{\hat{c}(P_j)}{\hat{q}(P_j)} \qquad (3.7.10)$$

is an unbiased estimator of I with respect to the random walk process $\{\hat{f}_n, \hat{p}_n\}$.

Proof. For convenience we shall denote

$$\hat{q}_j(P_1, \ldots, P_j) = \hat{q}(P_j), \qquad \hat{p}_k(P_1, \ldots, P_k) = \hat{p}(P_k), \qquad \text{etc.}$$

As in Section 3.6, we may write

$$E[\hat{I}] = \sum_{k=1}^{\infty} P[k(C) = k]E[\hat{I} \mid k(C) = k]$$

$$= \sum_{k=1}^{\infty} \int_\Gamma \cdots \int_\Gamma \frac{g(P_k)}{\hat{p}(P_k)} \prod_{j=1}^{k-1} \frac{\hat{c}(P_j)}{\hat{q}(P_j)} \prod_{j=2}^{k} \frac{K(P_j, P_{j-1})}{\hat{K}(P_j, P_{j-1})} \frac{S(P_1)}{\hat{S}(P_1)}$$

$$\times \prod_{j=2}^{k} \hat{K}(P_j, P_{j-1})\hat{S}(P_1) \prod_{j=1}^{k-1} \frac{\hat{q}(P_j)}{\hat{c}(P_j)} \hat{p}(P_k)\, dP_1 \cdots P_k$$

$$= \sum_{k=1}^{\infty} \int \cdots \int g(P_k)K(P_k, P_{k-1}) \cdots K(P_2, P_1)S(P_1)\, dP_1 \cdots dP_k$$

$$= I$$

by the L_1-convergence of the Neumann series and the boundedness of g.

Choosing $S = \hat{S}$, $K = \hat{K}$, $p = \hat{p}$ we see that

$$\hat{I}(C) = g(P_n)/p(P_n) \prod_{j=1}^{n-1} c(P_j)/q(P_j) \tag{3.7.11}$$

is an unbiased estimator of I with respect to the analog process. With

$$q(P) = c(P) = \frac{\Sigma_s(P)}{\Sigma_t(P)} \qquad \text{and} \qquad g(P) = \frac{\Sigma(P)}{\Sigma_t(P)},$$

we are led to the absorption estimator

$$\hat{I}(C) = \frac{\Sigma(P_n)}{\Sigma_a(P_n)}$$

discussed in Chapter 2, Eq. (2.5.10). The same proof shows the following.

Theorem 3.7 Let $\xi(P_1, \ldots, P_n)$ be any unbiased estimator of I with respect to the analog process. Then

$$\hat{\xi}(P_1, \ldots, P_n) = \xi(P_1, \ldots, P_n)X(P_1, \ldots, P_n),$$

where X is defined by Eq. (3.7.8), is an unbiased estimator of I with respect to the random walk process $\{\hat{f}_n, \hat{p}_n\}$.

We see that the choice

$$\xi(P_1, \ldots, P_n) = \frac{g(P_n)}{p(P_n)} \prod_{j=1}^{n-1} \frac{c(P_j)}{q(P_j)}$$

gives Theorem 3.6 as a special case of Theorem 3.7.

The conditions (3.7.7), although used in the proof of Theorem 3.7, are actually overly restrictive for some practical applications. For example, in the special nonanalog random walk process (Eq. 2.5.17) in which absorption is forbidden (at least for the first M collisions (cf. Eq. 3.7.9)) it is the case that $\hat{p}_j(P_1, \ldots, P_j) = 0$ even though $p(P_j) \neq 0$. This can lead to unbiased estimators even though it violates (3.7.7). The explanation lies in the fact that, even though the relation

$$\int_\Omega \xi \, d\mu = \int_\Omega \xi \frac{d\mu}{d\hat{\mu}} \, d\hat{\mu}$$

is not guaranteed for *all* ξ, it *is* satisfied for the specific estimator ξ involved in the Monte Carlo estimation. An example of this same situation arises in the next paragraph concerning zero-variance estimators. This discussion also serves to point out the need for care in deciding the circumstances under which a given estimator is unbiased.

We now show that one can choose $\hat{K}$ and $\hat{S}$ in such a way as to eliminate the variance in $\hat{I}$. To define the zero-variance process we need to know the importance function, that is, the solution of the equation

$$\psi^*(P) = \int K(P', P)\psi^*(P')\, dP' + g(P). \tag{3.7.12}$$

We have previously, in Section 3.6, shown that the function $\psi^*(P)$ may be interpreted as the importance of a neutron at P, as measured by the expected contribution of such a neutron to the integral I whose estimate is desired. If one assumes $\psi^*(P)$ is known, then one may define a random walk process $\{f_n^*, p_n^*\}$ by means of a source S^* and kernel K^* for a Fredholm equation as follows:

$$\begin{cases} S^*(P) = \dfrac{\psi^*(P)S(P)}{\int \psi^*(P)S(P)}, & K^*(P, P') = \dfrac{K(P, P')\psi^*(P)}{\psi^*(P')}, \\[2ex] \qquad\qquad p^*(P) = \dfrac{g(P)}{\psi^*(P)}. \end{cases} \tag{3.7.13}$$

Notice that

$$c^*(P) = \int_\Gamma K^*(Q, P)\, dQ = \int_\Gamma \frac{K(Q, P)\psi^*(Q)}{\psi^*(P)}\, dQ$$

$$= 1 - \frac{g(P)}{\psi^*(P)} = q^*(P) \leq 1.$$

Thus, the random walk process $\{f_n^*, p_n^*\}$ satisfies the conditions (3.3.18), (3.3.19), (3.7.9), which guarantee that it gives rise to a subcritical probability model. Even though condition (3.7.7) may be violated, one finds

Theorem 3.8 Let $\{f_n^*, p_n^*\}$ be defined by (3.7.13). Then the random variable (3.7.10) estimates I with zero variance.

Proof. One finds that, for every chain $C = \{P_1, \ldots, P_n\}$,

$$\hat{I}(C) = \frac{g(P_n)K(P_n, P_{n-1}) \cdots K(P_2, P_1)S(P_1)}{K^*(P_n, P_{n-1}) \cdots K^*(P_2, P_1)S^*(P_1)g(P_n)/\psi^*(P_n)}$$

$$= [\psi^*(P_1)/\psi^*(P_2)] \cdots [\psi^*(P_{n-1})/\psi^*(P_n)] \cdot \psi^*(P_n) \cdot \frac{\int \psi^*(P)S(P)\, dP}{\psi^*(P_1)}$$

$$= \int \psi^*(P)S(P)\, dP = I.$$

Since every chain gives the exact answer, the process must have zero variance.

We have already made use of one aspect of a duality theory for importance sampling, namely, that

$$I = \int g(P)\psi(P)\, dP = \int S(P)\psi^*(P)\, dP.$$

It is actually the case that a rather complete theory of duality exists for importance sampling which we should like to make explicit now. Despite the repetition, we once more write the integral equation for the collision density

$$\psi(P) = \int K(P, P')\psi(P')\, dP' + S(P) \tag{3.7.14}$$

and the integral

$$I = \int g(P)\psi(P)\, dP \tag{3.7.15}$$

to be estimated. The pair of equations

$$\psi^*(P) = \int K(P', P)\psi^*(P')\, dP' + g(P) \tag{3.7.16}$$

and

$$I = \int S(P)\psi^*(P)\, dP \tag{3.7.17}$$

has been shown to be adjoint to the pair (3.7.14) and (3.7.15), and vice versa. That is, with respect to a transport process defined by Eq. (3.7.16) and an integral (3.7.17) to be estimated, the function $\psi(P)$ is an importance function, which we call the *dual importance function*. It should now be clear that the theory previously developed for the pair (3.7.14), (3.7.15) applies equally well to the pair (3.7.16), (3.7.17). In particular, the estimator (3.7.10) dualizes to

$$I^*(C) = \frac{S(P_n)K(P_{n-1}, P_n) \cdots K(P_1, P_2)g(P_1)}{\hat{K}(P_n, P_{n-1}) \cdots \hat{K}(P_2, P_1)\hat{S}(P_1)\hat{p}(P_n)} \prod_{j=1}^{n-1} \frac{\hat{c}(P_j)}{\hat{q}(P_j)}, \tag{3.7.18}$$

which is an unbiased estimator of I with respect to the random walk process $\{f_n, p_n\}$ defined by $\hat{S}$, $\hat{K}$, and $\hat{p}$. The proof follows as did the proof of Theorem 3.6, since the series of integrals which represents the integral $\int S(P)\psi^*(P)\, dP$ is the same as that representing $\int g(P)\psi(P)\, dP$.

As in the proof of Theorem 3.8, one can show that, through proper choices of $\hat{K}$ and $\hat{S}$, the variance of $I^*(C)$ can be eliminated. Here, however, the zero-variance estimator involves the adjoint to ψ^*, i.e. the collision density $\psi(P)$. Assuming $\psi(P)$ known, the dual zero-variance process is obtained by dualizing Eqs. (3.7.13), as follows:

$$\begin{cases} S^*(P) = \dfrac{\psi(P)g(P)}{\int \psi(P)g(P)\, dP}, & K^*(P, P') = \dfrac{K(P', P)\psi(P)}{\psi(P')}, \\[3mm] \qquad\qquad p^*(P) = \dfrac{S(P)}{\psi(P)} \leq 1. \end{cases} \tag{3.7.19}$$

This gives the dual of Theorem 3.8, namely:

Theorem 3.9 With respect to the dual random walk process (f_n^D, p_n^D) defined by Eqs. (3.7.19), the random variable (3.7.18) estimates I with zero variance.

A theorem analogous to Theorem 3.7 also exists for dual importance sampling. To exhibit this one needs to construct an "analog" random walk process for sampling the adjoint equation. Let us assume that conditions (a), (b), and (c) hold for the adjoint kernel $K(P', P)$ and define

$$f_1(P) = \frac{g(P)}{\int_\Gamma g(P)\,dP}, \qquad f_n(P_1, \ldots, P_n) = f_n(P_2, \ldots, P_n \mid P_1) f_1(P_1)$$

$$= \prod_{l=2}^{n} \left[\frac{K(P_{l-1}, P_l)\,dP_l}{\int_\Gamma K(P_{l-1}, P_l)} \right] f_1(P_1). \qquad (3.7.20)$$

At this point p_k remains arbitrary, but it is sometimes convenient to take

$$p_k(P_k) = 1 - \int_\Gamma K(P_k, P)\,dP \qquad (3.7.21)$$

provided the integral on the right-hand side is positive and bounded by 1. It is easy, now, to establish

Theorem 3.10 If $\xi(P_1, \ldots, P_n)$ is an unbiased estimator of I with respect to the analog process (3.7.20), (3.7.21) for the adjoint equation, then

$$\hat{\xi}(P_1, \ldots, P_n) = \xi(P_1, \ldots, P_n) X(P_1, \ldots, P_n) \qquad (3.7.22)$$

is an unbiased estimator of I with respect to the random walk process defined by $\hat{S}$, $\hat{K}$, $\hat{p}$.

The preceding theory has depended on certain natural assumptions (conditions (a) and (b) of Section 3.3)† made on the kernel K and on the assumption of the boundedness of the function g. By this time it should be apparent that we might have worked as well with the kernel L for the integral equation

$$\chi(P) = \int L(P, P') \chi(P')\,dP' + Q(P) \qquad (3.7.23)$$

and that a similar duality holds for equations such as (3.7.23). Once again we wish to point out that the difference is that the source Q is a physical source, while the source S to the collision density is a density of first collisions. Furthermore, the kernel $L(P, Q)$ gives the expected density of particles coming out of collision at P given that one has come out of collision (or source) at Q, while $K(P, Q)$ gives the expected densities entering collision. Since K and L describe the same physical process, the subcriticality‡ of one implies that of the other and the preceding theorems apply equally well to either.

---

† Of course, the theory is valid, also, under the weaker condition that $\mathscr{K}$ be sub-critical (see discussion, p. 94).

‡ In the integral operator sense defined on p. 94.

We have seen that the kernel conditions guarantee that the analog process be reasonable in the sense that infinite chains occur with probability zero, and also give the L_1-convergence of the Neumann series. This convergence, together with the boundedness of g, implies the convergence of the series of integrals which represents I. We should now like to relax the assumption that g be bounded to see whether we can obtain estimates of the solution $\psi(P_0)$ at a single point by effectively letting $g(P)$ tend to $\delta(P - P_0)$. Suppose, e.g., that we assume that the Neumann series converges at P_0 to $\psi(P_0)$. Then Theorem 3.6, though it remains valid, would lead to an impractical estimator. It is easy to see that the contribution of each chain would be zero or infinite and the resulting random variable can be shown to have infinite variance. One would also rapidly discover that the application of Theorem 3.7 would suffer from the same defect for all reasonable choices of ξ.

The dual estimator (3.7.18), with $\hat{S}(P) = g(P) = \delta(P - P_0)$, does not have this drawback. However, in examining Eqs. (3.7.16) and (3.7.17) we see that the estimation of I by simulating the ψ^*-equation involves the knowledge of the function S, the first collision density. Since this function may be difficult to write in closed form, this technique based on (3.7.16) suffers from this defect. If, however, we make use of the integral equation for χ,

$$\chi(P) = \int L(P, P')\chi(P')\, dP' + Q(P) \qquad (3.7.24)$$

and the integral

$$I = \iiint g(\mathbf{r}', \mathbf{E})T(\mathbf{r}, \mathbf{r}'; \mathbf{E})\chi(\mathbf{r}, \mathbf{E})\, d\mathbf{r}'\, d\mathbf{r}\, d\mathbf{E}, \qquad (3.7.25)$$

then an appropriate adjoint equation is

$$\chi^*(P) = \int L(P', P)\chi^*(P')\, dP' + \int g(\mathbf{r}', \mathbf{E})T(\mathbf{r}, \mathbf{r}'; \mathbf{E})\, d\mathbf{r}' \qquad (3.7.26)$$

and we have

$$I = \int_\Gamma Q(P)\chi^*(P)\, dP. \qquad (3.7.27)$$

Now, the integral I of Eq. (3.7.27) involves the physical source density Q rather than S. Furthermore, the choice $g(\mathbf{r}', \mathbf{E}) = \delta(\mathbf{r}' - \mathbf{r}_0)\delta(\mathbf{E} - \mathbf{E}_0)$ leads to a perfectly reasonable adjoint process which may be used to construct estimates of χ^* and, thereby, through (3.7.27), estimates of $I = \psi(\mathbf{r}_0, \mathbf{E}_0)$. Use is made of this method in the application of Chapter 5, except that the integro-differential transport equation is used in the derivation. We have already alluded to the estimator (3.6.22) developed by Kalos (Ref. 17) for an alternative method of estimating the collision density at a point due to a point source.

The Eq. (3.7.26) may be shown to be identical with that obtained by taking the adjoint of the integro-differential equation for the directional flux, as in Chapter 5 (cf. Eq. 5.3.6). Under the conditions of Theorem 3.4, this enables us to identify χ^* with the adjoint flux F^*, even though $\chi \neq F$. This latter fact is not surprising in view of the fact that F^* has been obtained by taking the adjoint of an integro-differential operator while χ^* arises as the adjoint of an integral operator. Thus, while χ is the source to the equation for F (see Eq. 2.4.23), χ^* is not the source to the equation for F^*.

A second closely related method of estimating $\psi(P_0)$ is actually due originally to Albert (Ref. 12). We shall indicate how Albert's estimator arises naturally within the context of dual importance sampling. Suppose, then, that we want to estimate $\psi(P_0)$, the collision density at P_0. From the transport equation

$$\psi(P_0) - S(P_0) = \int_\Gamma K(P_0, P)\psi(P) \, dP. \tag{3.7.28}$$

Thus, we may apply the theory developed previously to estimate the integral (3.7.28), with $g(P) = K(P_0, P)$. The estimator (3.7.18) applied here yields the random variable

$$I^*(C) = \frac{S(P_n)K(P_{n-1}, P_n) \cdots K(P_1, P_2)K(P_0, P_1)}{\hat{K}(P_n, P_{n-1}) \cdots \hat{K}(P_2, P_1)\hat{S}(P_1)\hat{p}(P_n)} \prod_{j=1}^{n-1} \frac{\hat{c}(P_j)}{\hat{q}(P_j)}, \tag{3.7.29}$$

which can be shown to give an unbiased estimate of $\psi(P_0) - S(P_0)$ (even though $g(P) = K(P_0, P)$ may *not* be bounded) by the assumed convergence of the Neumann series to ψ at $P = P_0$. Thus, the random variable

$$I_0(C) = S(P_0) + I^*(C) \tag{3.7.30}$$

gives an unbiased estimator of $\psi(P_0)$ which is, in fact, identical with Albert's estimator. This estimator is interesting in that it does not necessarily require particles to start at P_0.

If, in Eq. (3.7.18), we make the choices

$$g(P_1) = \hat{S}(P_1) = \delta(P_1 - P_0), \qquad \hat{K}(P_l, P_{l-1}) = K(P_{l-1}, P_l),$$

then the random walk process which results is like an adjoint simulation in which all particles originate at P_0. They also make their first collisions at P_0 because of the choice $\hat{S}(P_1) = \delta(P_1 - P_0)$. The distinction between the states P_0 and P_1 in this simulation process is that P_0 is not a real collision point, so that termination may not occur at P_0. P_1, on the other hand, denotes the first collision point and, because of this interpretation, we allow the possibility of termination at P_1. The estimator (3.7.18) becomes

$$I^*(C) = \frac{S(P_n)}{\hat{p}(P_n)} \prod_{j=1}^{n-1} \frac{\hat{c}(P_j)}{\hat{q}(P_j)}, \qquad n \geq 1. \tag{3.7.31}$$

Notice that for $n = 1$ this gives $S(P_1)/\hat{p}(P_1)$, which has expected value

$$E_1 = \int \frac{S(P_1)}{\hat{p}(P_1)} \hat{p}(P_1) \, \delta(P_1 - P_0) \, dP_1 = S(P_0).$$

For $n = 2$, (3.7.31) has the expected value $E_2 = \int K(P_0, P_2)S(P_2) \, dP_2$; and for $n > 2$, (3.7.31) has the expected value

$$E_n = \int K(P_0, P_2)K(P_2, P_3) \cdots K(P_{n-1}, P_n)S(P_n) \, dP_n \, dP_{n-1} \cdots dP_2.$$

We notice that these terms form an infinite series representation of $\psi(P_0)$.

If, on the other hand, in Eq. (3.7.29) we choose

$$\hat{S}(P_1) = \frac{K(P_0, P_1)}{\int K(P_0, P_1) \, dP_1}, \qquad \hat{K}(P_l, P_{l-1}) = K(P_{l-1}, P_l),$$

then $\hat{c}(P_0) = \int K(P_0, P_1) \, dP_1$ and (3.7.29) becomes

$$I^*(C) = \frac{S(P_n)\hat{c}(P_0)}{\hat{p}(P_n)} \prod_{j=1}^{n-1} \frac{\hat{c}(P_j)}{\hat{q}(P_j)}, \qquad n \geq 1. \tag{3.7.31a}$$

For $n = 1$, (3.7.31a) has expected value

$$E_1' = \int \frac{S(P_1)\hat{c}(P_0)}{\hat{p}(P_1)} \frac{K(P_0, P_1)}{\hat{c}(P_0)} \hat{p}(P_1) \, dP_1 = \int K(P_0, P_1)S(P_1) \, dP_1,$$

while for $n \geq 2$, (3.7.31a) has expected value

$$E_n' = \int K(P_0, P_1)K(P_1, P_2) \cdots K(P_{n-1}, P_n)S(P_n) \, dP_n \cdots dP_1.$$

We see that $E_n' = E_{n+1}$, $n \geq 1$; thus, the estimator (3.7.31a) accounts for all contributions to $\psi(P_0)$ except for $S(P_0)$. This is, of course, to be expected, since (3.7.29) is supposed to give an unbiased estimate of $\psi(P_0) - S(P_0)$. Thus, $I_0(C)$ is seen to be a true adjoint estimator of the collision density at P_0, which, because of the correspondence just established between (3.7.31) and (3.7.31a), may be thought of as arising by starting all particles at P_0.

Thus we have shown that a special case of Albert's estimator is equivalent to an estimator obtained by solving the adjoint transport equation. We shall become much more explicit in Chapter 5 in describing the estimation of flux and activation rates at a point by solving adjoint equations.

The material we have covered so far in Section 3.7 is quite general, perhaps too general in the sense that it is not easy to see how the ideas may be applied directly. To remedy this situation we shall derive, as a special case of our formalism, the exponential transformation. The latter is a variance-reducing device which has been exploited with some success in shielding applications in one-dimensional arrays (see Refs. 8–10, 24–27). We shall follow the work of Leimdörfer (Refs. 8, 10, 25, 26) and Kalos (Ref. 22) in our derivation.

If one examines Eqs. (3.7.13) which define the zero-variance random walk process $\{f_n^*, p_n^*\}$ in terms of the importance function $\psi^*(P)$, one sees that approximations to the zero-variance process may be defined in terms of approximations to the importance function. For concreteness, we shall work with the equation

$$\chi(P) = \int L(P, P')\chi(P')\, dP' + Q(P) \tag{3.7.32}$$

for the density χ of particles coming out of collisions. We assume the integral

$$I = \int_\Gamma h(P)\chi(P)\, dP \tag{3.7.33}$$

is to be estimated. We take a point of view closely related but inverse to that used before in deriving an equation for a weighted density of collisions. Let $I(P)$ denote a function which is to be used to modify the source Q and kernel L in such a way that "important" events are emphasized. $I(P)$ in some sense represents an approximation to the solution to the adjoint equation and is, thus, an approximate importance function. If we multiply (3.7.32) by

$$\frac{I(P)}{\int_\Gamma I(P)Q(P)\, dP}$$

we get

$$\hat{\chi}(P) = \int_\Gamma \hat{L}(P, P')\hat{\chi}(P')\, dP' + \hat{Q}(P), \tag{3.7.34}$$

where

$$\hat{Q}(P) = \frac{I(P)Q(P)}{\int_\Gamma I(P)Q(P)\, dP}, \qquad \hat{L}(P, P') = \frac{I(P)}{I(P')} L(P, P'),$$

$$\hat{\chi}(P) = \frac{\chi(P)I(P)}{\int_\Gamma I(P)Q(P)\, dP} = \frac{\chi(P)I(P)}{N}, \qquad N = \int_\Gamma I(P)Q(P)\, dP,$$

so that

$$I = N \int_\Gamma \frac{h(P)}{I(P)} \hat{\chi}(P)\, dP. \tag{3.7.35}$$

Thus (in principle, at least), methods previously developed for estimating integrals may be applied to the estimation of (3.7.35) applied to the transformed density $\hat{\chi}(P)$. In practice we find that $I(P)$ must be a rather simple function, otherwise the normalizing integral N may be difficult to calculate. N is, after all, an approximation to the integral I and is exact if $I(P) = \chi^*(P)$ is the solution to the adjoint equation. Furthermore, it may be difficult to evaluate $\hat{c}(P) = \int_\Gamma \hat{L}(Q, P) \, dQ$ for the transformed kernel. Finally, as we shall see, weight fluctuations must be handled in such a fashion that the variance is not greatly increased by them.

Suppose the source $Q(z, E, \mu)$ is confined to a region near $z = 0$ in a one-dimensional problem, where $\mu = \cos^{-1} \theta$ and θ is the angle between the positive z-axis and the direction of travel. If one wants estimates of reaction rates for large z, an analog simulation based on Eq. (3.7.32) will result in poor statistics, since the flux will decay approximately exponentially with z and not many particles will survive to deep penetrations. One alternative is to use the expected value estimators developed in Section 3.6 which permit contributions to I from particles at only moderate penetrations. One might also, as an alternative, alter the source and kernel in such a way that in Eq. (3.7.34) the number of particles is roughly independent of z.

Based on these considerations, the exponential transformation involves the choice $I(z, E, \mu) = \exp(cz)$. For the moment, c may be arbitrary, but we shall presently see that certain restrictions on c are desirable.

To make use of Eq. (3.7.34), one may sample $Q(z, E, \mu)$, obtain source particles, and weight each particle by

$$\frac{\exp(cz)}{\iiint \exp(cz)Q(z, E, \mu) \, dz \, dE \, d\mu},$$

or try to sample $\hat{Q}$ directly. Similarly, one may use the kernel L to determine flight from z to z' and then multiply by $\exp[c(z' - z)]$. Finally, in scoring procedures, one uses $Nh(P)/I(P)$ in place of $h(P)$ to determine the final weighted contribution to the integral. However, the advantage of using Eq. (3.7.34) will be lost unless one can actually sample $\hat{Q}$ (or a density close to it) directly. Similarly, one must be able to move particles from point to point in phase space by means of a kernel closely related to $\hat{L}$. It is easy to see that, unless this is accomplished, nothing will be gained (in terms of variance) by simulating Eq. (3.7.34) instead of the original Eq. (3.7.32).

The usual procedure is to incorporate the factor $\exp(cz)$ into the transport kernel T (cf. Eq. 2.4.9). The most natural way of doing so is to define a new transport kernel T^* by (2.4.9) but with $\Sigma_t^* = \Sigma_t - \mu c$ replacing Σ_t. Then, if T^* is used to determine intercollision distances, the weight must be modified by the factor

$$T/T^* = \frac{\Sigma_t \exp(-\Sigma_t z)}{\Sigma_t^* \exp(-\Sigma_t^* z)} = \frac{\Sigma_t}{\Sigma_t^*} \exp(-\mu c z). \tag{3.7.36}$$

Notice that care must be taken that $\Sigma_t^* \neq 0$. This is usually accomplished by taking $0 < c < \Sigma_t$. With c so chosen Σ_t^* will be smaller for positive μ than for negative μ, so that longer forward flights are enhanced and backward motion is inhibited. References 8 and 10 give a good discussion of how c may be chosen to give good results and, in addition, describe some methods for using negative Σ_t^*. Fluctuating weights are usually controlled to some extent by the use of Russian roulette and splitting, devices which are discussed in the next section.

3.8 RUSSIAN ROULETTE AND SPLITTING

The methods of this section are two fairly well-known Monte Carlo techniques for reducing variance, techniques which cannot easily be classified as "analog" or "nonanalog." We shall first attempt to motivate both procedures and then define them formally.

In any analog simulation of a transport problem one is strongly tempted to divide the phase space into regions of two broad types, which might be called important or unimportant. In the unimportant regions one wants to terminate a high percentage of random walk chains, either because the probability of contribution from such regions is low, or because such chains are atypical and may thus add greatly to the variance of the process. In important regions the exactly opposite conditions hold: in such regions one wants to magnify the number of samples. One also wants to encourage particles to migrate from unimportant to more important regions. The methods of importance sampling of Section 3.7 are designed to achieve this, but some knowledge of the importance function or dual importance function is needed there. Furthermore, use of a poor approximation to the importance function may actually result in a worsening of the statistics—i.e. a larger variance than if no importance sampling is employed. Of course, to some extent the methods of the present section suffer from the same potential defect and must be used with care if real increases in efficiency are to result.

Suppose one attempts something somewhat less ambitious than is the goal in importance sampling—namely, to play a game of Russian roulette with particles which have entered "bad" regions. As the name implies, such a game is rigged to terminate a high percentage of such random walk histories, but attaches an excess weight to the survivors, so that the resulting estimator remains unbiased. In this context Russian roulette is a special case of importance sampling in which the probabilities of termination have been made high in bad regions and low in good ones. The splitting is an additional technique, used in good regions, whose effect is to split each chain which arrives in such a region into n independent branches, each having a weight which is $1/n$ times the weight of the parent chain.† The combination of these two techniques can be extremely effective in achieving large reductions in variance, particularly when reasonably good notions of important and unimportant regions are easily found, such as, for example, in shielding calculations.

---

† Actually, it is only necessary that the total weight of the progeny be the same as the weight of the parent chain.

To define a Russian roulette estimator formally, we first specialize to a particular random walk process $\{f_n, p_n\}$. For the transition probabilities f_n we choose the analog functions as defined by Eqs. (3.3.8) and (3.3.9), i.e.

$$f_n(P_1, \ldots, P_n) = \left[\prod_{l=2}^{n} \frac{K(P_l, P_{l-1})}{c(P_{l-1})} \right] S(P_1), f_1(P_1) = S(P_1), \quad (3.8.1)$$

where

$$c(P) = \int_\Gamma K(Q, P) \, dQ,$$

but for the termination probabilities p_n we do not take the analog ones, $p_n(P) = \Sigma_a(P)/\Sigma_t(P)$. In fact, the whole point of Russian roulette is to magnify these numbers p_n in bad regions; thus, they should be chosen to be significantly larger than Σ_a/Σ_t in such regions. At any rate, with f_n defined by (3.8.1) and with p_n arbitrary but subject to (3.7.7), an estimator for Russian roulette is (3.7.10), which simplifies in this case to

$$\hat{I}(C) = \frac{g(P_n)}{p_n(P_n)} \prod_{j=1}^{n-1} \frac{c(P_j)}{q(P_j)}. \quad (3.8.2)$$

The estimator (3.8.2) is certainly unbiased, being a special case of (3.7.10), and one may readily see how (3.8.2) accomplishes the desired effect. Suppose, e.g., that

$$\int K(P, Q) \, dP = \Sigma_s(Q)/\Sigma_t(Q) \le 1$$

as in a non-multiplying medium and one wishes to terminate, say, 90% of all histories in a certain region B. Then we choose

$$p(P) = 1 - q(P) = 0.9 \qquad \text{for } P \in B$$

and the effect is to introduce in (3.8.2) a factor

$$\frac{c(P_i)}{0.1} = \frac{\int K(P, P_i) \, dP}{0.1} = 10 \int K(P, P_i) \, dP$$

$$= 10 \Sigma_s(P_i)/\Sigma_t(P_i), \qquad \left(\text{if } c(P_i) = \frac{\Sigma_s(P_i)}{\Sigma_t(P_i)} \right),$$

at each collision point $P_i \in B$, if this collision is survived. The quantity $10 \Sigma_s(P_i)/\Sigma_t(P_i)$ is the excess weight referred to earlier. In general, the choice

$$q(P_i) = \frac{1}{n} \int K(P, P_i) \, dP, \qquad P_i \in B$$

reduces the probability of survival in B by a factor of n but multiplies the weights of survivors by a factor of n. Now suppose we choose $q(P) = 1$ for all P so that every particle survives every collision. Then a multiplicative factor of $c(P)$ must modify the random variable upon collision at P. This gives rise to the nonanalog process discussed in Section 2.5 (Eq. 2.5.17). However, if $q(P) = 1$ for *every* P, then condition (3.7.8) and, hence, $\mu(\Lambda_\infty) = 0$, will clearly be violated. In practice, after a fixed number of collisions are performed or after the weight of the particle falls below some minimum, the choice $q(P) < 1$ is made in such a way that condition (3.7.8) is satisfied. This ensures that the probability model will be subcritical, as desired.

The splitting technique is somewhat more complicated to define formally. Let G be a region where splitting is to take place. For simplicity, we consider only the case where each chain splits into exactly two independent chains upon collision in G. The results easily generalize to a finite number of splitting regions $G_1, \ldots, G_r$ with n_i independent branches in region G_i, $i = 1, \ldots, r$.

We shall assume an arbitrary random walk process $\{f_n, p_n\}$, although in actual applications one usually begins with the analog process. Each time an i-th collision is made at a point $P_i \in G$ and the chain is not terminated at P_i, the chain is split into two independent branches stemming from P_i by choosing two independent points P_{i+1}, P'_{i+1} from the conditional density function $f_{i+1}(Q_{i+1} \mid P_1, \ldots, P_i)$. This process is repeated every time a new point of G is encountered. The result is that, from an initial chain point P_1 chosen from $f_1(P_1)$, a possibly large number of chains may result, each with initial chain point P_1, through the process of splitting. Let $G(P_1)$ denote the collection of all chains stemming from an initial point P_1.

We shall denote by $C^{(k)}$ a chain which has suffered exactly k splittings and let m_k be the number of chains of $G(P_1)$ which have split exactly k times, $k = 0, 1, \ldots$ A chain $C^{(k)}$ is then one which has suffered exactly k collisions in G which did not result in termination. In the analog process, since chains of infinite length have probability zero by Theorem 3.3, the total number of chains which can originate from a given initial point P_1 through this splitting process is finite with probability one. Thus, we can enumerate the elements of $G(P_1)$ as

$$G(P_1) = \{C_{i_0}^{(0)}, \ldots, C_{i_K}^{(K)}\}, \qquad (3.8.3)$$

where the subscript i_j ranges from 1 to m_j (if $m_j = 0$, there are no chains which have split exactly j times). Thus, K is the largest number of splittings undergone by a single chain originating at P_1 and is finite with probability one for the analog process. The appropriate random variable† to associate

---

† The astute reader may notice that the use of (3.8.4) is slightly different from, though entirely equivalent to, the more usual way of describing splitting. The usual treatment involves a branching process where the number of progeny continually increases as more and more splittings occur. Our treatment views the splitting not really as a creative process but as one which merely separates fragments at certain collisions. Prior to the splitting collisions, the fragments are then regarded as coalesced into a single particle of larger weight.

with the set $G(P_1)$ is

$$\xi\{G(P_1)\} = \sum_{k=0}^{K} \sum_{i_k=1}^{m_k} \frac{1}{2^k} \xi[C_{i_k}^{(k)}], \tag{3.8.4}$$

where ξ is any unbiased analog random variable. It is understood that if any $m_k = 0$, then that term is omitted from the sum on the right.

The expected value of (3.8.4) is

$$
\begin{aligned}
E[\xi\{G(P_1)\}] &= \sum_{k=0}^{K} \sum_{i_k=1}^{m_k} \frac{1}{2^k} E[\xi\{C_{i_k}^{(k)}\}] \\
&= \sum_{k=0}^{K} \frac{m_k}{2^k} E[\xi\{C_{i_k}^{(k)}\}] \\
&= \sum_{k=0}^{K} \frac{m_k}{2^k} I_k(P_1),
\end{aligned}
$$

where $I_k(P_1)$ is the conditional expected value of ξ restricted to chains which originate at P_1 and have split exactly k times.

To demonstrate that $\xi\{G(P_1)\}$ is an unbiased estimator of I we show that

$$\sum_{k=0}^{K} \frac{m_k}{2^k} I_k(P_1) = I(P_1),$$

as follows.† Each chain in $G(P_1)$ may be regarded as a separate history. If there are $N(= \sum_{k=0}^{K} m_k)$ such chains in $G(P_1)$ then we have, through the splitting process, generated N histories but, in a sense, these N histories are not all equally probable. For suppose that a particle starts at P_1 and is forced to select, at random, any one of the N chains. At each splitting point the particle can, with equal probability, move in either of two directions. Therefore, the probability, p_{i_k}, that it will stay on any particular k-chain, $C_{i_k}^{(k)}$, is equal to $1/2^k$. Thus the estimator defined through (3.8.4) is an average of the value of ξ on each chain, correctly weighted by the probability p_{i_k} of that chain. Bayes' Theorem then yields the desired result.

We note that since $1/2^k$ is the probability of a k-chain, $C_{i_k}^{(k)}$, and since the particle must select one of the N chains, the sum

$$\sum_{k=0}^{K} \frac{m_k}{2^k}$$

must be equal to one. This may be proved by a more formal argument, as follows.

---

† The reader should understand that

$$I = \int I(P_1) S(P_1) \, dP_1.$$

Theorem 3.11 In the splitting process defined above

$$\sum_{k=0}^{K} \frac{m_k}{2^k} = 1, \tag{3.8.5}$$

where m_k is the number of chains which have split exactly k times.

Proof. The proof is by induction on K, the largest number of splittings. If $K = 0$, then necessarily $m_0 = 1$ and $m_k = 0$ for $k > 0$. Thus, the theorem is trivially true for $K = 0$. Now assume the conclusion of the theorem for $K > 0$ and we shall prove it for $K + 1$. The induction assumption applies to any collection consisting of chains which have split at most K times. Now consider a collection $G(P_1)$ containing chains which have split $(K + 1)$ or fewer times. For such a collection it is clear that m_{K+1} is even, since exactly two chains were produced on each $(K + 1)$-th split. We write $m_{K+1} = 2N$. Then N is the total number of additional K-chains that would have resulted if none of the $(K + 1)$-th splittings had occurred, i.e. if each K-chain had terminated before it had a chance to split again. Thus, if there were m_K chains which had split exactly K times (not counting the N which split again) in the original $G(P_1)$, there would be $(m_K + N)$ such K-chains in the new collection resulting upon termination of each K-chain before it could split again. Denote this last collection by $G'(P_1)$. Then for $G'(P_1)$ we can write

$$\sum_{k=0}^{K-1} \frac{m_k}{2^k} + \frac{m_K + N}{2^K} = 1.$$

But since $m_{K+1} = 2N$, this sum is also

$$\sum_{k=0}^{K-1} \frac{m_k}{2^k} + \frac{m_K}{2^K} + \frac{m_{K+1}}{2^{K+1}} = \sum_{k=0}^{K+1} \frac{m_k}{2^k},$$

which completes the proof.

If the random variable ξ of (3.8.4) is the estimator (3.8.2) of Russian roulette, then (3.8.4) involves both the techniques of Russian roulette and splitting at the same time. As we have shown by our discussion, they may be used independently but are often combined in the manner previously suggested. By emphasizing regions of importance by splitting and suppressing regions of unimportance by Russian roulette, some of the effect of importance sampling is achieved but without the necessity of altering transition probabilities.

It is interesting to note (as have Goertzel and Kalos in Ref. 21) that the exponential transformation may be obtained from splitting and Russian roulette by a limiting argument. Again, in a one-dimensional case, let the points $z = a_i$ ($i = 1, 2, \dots, n$) define a set of splitting boundaries. When a particle crosses any such boundary moving to the right, we split the particle into an average of $\rho > 1$ particles, each carrying the weight of ρ^{-1} times its original weight. When a particle crosses such a boundary moving to the left, Russian roulette is used to terminate the history with probability ρ^{-1}; if the particle survives, its weight is multiplied by ρ.

One can show that the introduction of this roulette and splitting scheme is entirely equivalent to the following modification of the transport kernel T:

$$T^*(z', z; E, \mu) = \rho^{\sum_{i=1}^{n}[\eta(z-a_i)-\eta(z'-a_i)]} T(z', z; E, \mu), \qquad (3.8.6)$$

where

$$\eta(x) = \begin{cases} 1 & x > 0, \\ 0 & x \le 0, \end{cases}$$

and T is the ordinary transport kernel.

If we take $\rho = 1 + (1/n)$ and $a_i = (i/n)D$, where D is the total thickness to be considered, then, for n sufficiently large, the continuous variable z may be thought of as approximated by the discrete variable $(i/n)D$, so $i \approx nz/D$. Thus,

$$\sum_{i=1}^{n} \eta(z - a_i) = \frac{nz}{D}.$$

Taking the limit as $n \to \infty$ gives

$$\lim_{n \to \infty} \rho^{\sum_{i=1}^{n}[\eta(z-a_i)-\eta(z'-a_i)]} = \lim_{n \to \infty} \left(1 + \frac{1}{n}\right)^{\alpha n(z-z')}$$

$$= e^{\alpha(z-z')}, \qquad \alpha \equiv \frac{1}{D},$$

which corresponds to the exponential transform $I(z, E, \mu) = e^{\alpha z}$ (cf. discussion on p. 123).

3.9 CORRELATED SAMPLING

We shall use the term "correlated sampling" in referring to the performance of a pair of Monte Carlo calculations whose associated random variables have strong positive correlation. The method is especially useful in calculating differential effects by Monte Carlo. In such cases it is typical that one wants to calculate a small change in a system due to a perturbation of some sort (e.g. a Doppler coefficient of temperature, a change in thermal utilization due to a change in atom density, etc.). If one were to undertake to calculate such changes by performing two independent Monte Carlo runs and then observing the perturbation, it would often be the case that the statistical uncertainty in the perturbation would be large compared with the perturbation. To circumvent this difficulty, it is essential to correlate the two runs positively so that, to as great an extent as is possible, only the effects of the perturbation itself are subject to statistical fluctuation.

A simple, yet fairly effective, way of achieving positive correlation is through control of the random numbers in the two problems. Thus, two problems are still run—the base problem and the perturbed problem—but some of the histories in both will suffer the same fate through control of random numbers.

A useful method of controlling the random numbers may be described in the following way. Suppose the algorithm adopted for generating pseudo-random numbers gives the sequence $\rho_1, \rho_2, \ldots$, of numbers. By a rearrangement of the sequence we may guarantee that all histories in the base and perturbed problems begin with the same pseudo-random number without altering the statistical randomness of the sequence. If we use double subscripts i, j to convey the idea that pseudo-random number ρ_{ij} is the j-th pseudo-random number used for Monte Carlo history i and use the symbol $S(\rho)$ to denote the successor to ρ in the original sequential ordering, then we define

$$\rho_{1,1} = \rho_1, \qquad \rho_{1,2} = \rho_2, \qquad \rho_{1,3} = \rho_3, \ldots, \qquad \rho_{1,n} = \rho_n$$
$$\rho_{2,1} = 1 - \rho_{1,2}, \qquad \rho_{2,2} = S(\rho_{2,1}), \ldots \qquad\qquad (3.9.1)$$
$$\rho_{3,1} = 1 - \rho_{2,2}, \qquad \rho_{3,2} = S(\rho_{3,1}), \ldots$$

$$\cdot$$
$$\cdot$$
$$\cdot$$

Since $(1 - x)$ and x have the same distribution if x is uniform on $0 \leq x \leq 1$, this method destroys none of the statistical properties of the sequences but achieves the desired effect of starting the same numbered histories with the same pseudo-random number. The only precaution to be taken is that each history must use at least two pseudo-random numbers—this is so in all but the most trivial cases. Even if some history, say history n, used only a single random number, one could still generate $S(\rho_{n,1})$ without using it for history n, but only to obtain $\rho_{n+1,1}$. The above-described technique has been incorporated into the MARC thermal multigroup program to calculate differential effects. Some of these results are reported in Ref. 28.

It is understood that when correlated sampling is used, statistical analysis is performed on the differential effects unless the amount of correlation can be calculated and used directly in a formula for variance of the desired perturbation.

A potentially more powerful way of achieving correlation between two problems is to combine the two into a single problem, using a single random walk process $\{f_n, p_n\}$ to generate particle histories but using different random variables for each of the two problems. This method, when it is applicable, has the further advantage of saving computing time, since only a single set of histories is processed. The method has been used in many programs such as TRIGR-P and TRIGR-S (Ref. 29), where differential effects resulting from geometry changes are treated. Olhoeft (Ref. 30) has developed a method for calculating the Doppler coefficient by obtaining a Neumann series expression for the derivative and then applying fairly standard methods to its estimation. Thus, his technique is different from the one we shall discuss now but is still aimed at computing differential effects.

Suppose, then, that the two problems in question may be described, respectively, by two integral equations for the collision densities:

$$\psi(P) = \int K(P, P')\psi(P')\, dP' + S(P) \tag{3.9.2}$$

and

$$\hat{\psi}(P) = \int \hat{K}(P, P')\hat{\psi}(P')\, dP' + \hat{S}(P), \tag{3.9.3}$$

where, in some sense, S and $\hat{S}$ are close and K and $\hat{K}$ are close. Suppose further it is desired to estimate

$$I = \int_{\Gamma} g(P)[\psi(P) - \hat{\psi}(P)]\, dP, \tag{3.9.4}$$

where $g(P)$ is a known, bounded function. Let $\{f_n, p_n\}$, $\{\hat{f}_n, \hat{p}_n\}$ denote the analog processes for Eqs. (3.9.2), (3.9.3), respectively, as defined by Eqs. (3.3.8)–(3.3.10). Just as in Section 3.7 (cf. Eq. 3.7.7) some care must be exercised as to which random walk process is used to define the base problem. Suppose conditions (3.7.6) are satisfied: assume, further, that the function X, defined by Eq. (3.7.8) is bounded except possibly for sets of chains having probability zero in the $\hat{\mu}$-process. Then the process (f_n, p_n) may be taken to define the basic random walk and we may write

$$X(P_1, \ldots, P_k) = \frac{d\mu}{d\hat{\mu}}(P_1, \ldots, P_k). \tag{3.9.5}$$

Now we may make direct use of Theorem 3.7 to define an estimator for I. We postulate that ξ is any unbiased estimator of $\int g(P)\psi(P)\, dP$ with respect to the analog process $\{f_n, p_n\}$. Then

$$\hat{\xi}(P_1, \ldots, P_n) = \xi(P_1, \ldots, P_n)X(P_1, \ldots, P_n)$$

is also an unbiased estimator of $\int g(P)\psi(P)\, dP$ with respect to the process $\{\hat{f}_n, \hat{p}_n\}$. Now let ξ^* be the estimator of $\int g(P)\hat{\psi}(P)\, dP$ which corresponds (in the process $\{\hat{f}_n, \hat{p}_n\}$ to the estimator ξ. Then ξ^* and ξ are actually identical when considered as functions on sequences of collision points. The theorem on correlated sampling based on these ideas is as follows.

Theorem 3.12 The random variable $\hat{\xi} - \xi^*$ is an unbiased estimator of $I = \int_{\Gamma} g(P)[\psi(P) - \hat{\psi}(P)]\, dP$ with respect to the random walk process $\{\hat{f}_n, \hat{p}_n\}$.

The proof is an immediate consequence of Theorem 3.7, since

$$E[\hat{\xi} - \xi^*] = E[\hat{\xi}] - E[\xi^*] = \int_\Gamma g(P)\psi(P)\,dP - \int_\Gamma g(P)\hat{\psi}(P)\,dP.$$

The significance of the result may now be seen. If the two random walk processes are "close," then $\hat{\xi}$ and ξ are nearly equal, so $\hat{\xi}$ and ξ^* will be nearly equal. More important, $\hat{\xi}$ and ξ^* will be strongly positively correlated, so that the variance in their difference,

$$V[\hat{\xi} - \xi^*] = V[\hat{\xi}] + V[\xi^*] - 2 \operatorname{cov}[\hat{\xi}, \xi^*], \qquad (3.9.6)$$

may be small, significantly smaller than $V[\hat{\xi}] + V[\xi^*]$, which would be the variance in the difference in the absence of any correlation. Since

$$\operatorname{cov}[\hat{\xi}, \xi^*] = E\{(\hat{\xi} - E[\xi])(\xi^* - E[\xi^*])\}$$
$$= E[\hat{\xi} \cdot \xi^*] - E[\hat{\xi}]E[\xi^*] \qquad (3.9.7)$$

we observe that if the two analog processes do, in fact, exactly coincide, then $\hat{\xi} = \xi^* = \xi$ and (3.9.6) gives

$$V[\xi - \xi^*] = 2V[\xi] - 2 \operatorname{cov}(\xi, \xi)$$
$$= 2V[\xi] - 2V[\xi] = 0,$$

so that the true mean, which is zero, is being estimated with no variance, as one would hope.

3.10 ANTITHETIC VARIATES

In Section 3.9 advantageous use was made of a strong positive correlation between two random variables. In this section we seek negatively correlated variables.

The method of antithetic variates was first developed by Hammersley and Morton (Ref. 31) in connection with the estimation of definite integrals, and has been expanded on by several authors, notably Hammersley and Handscomb in Ref. 1, and Hammersley and Mauldon in Ref. 32. In Ref. 1, a general theorem (proved in Ref. 33) concerning this method is stated and the ramifications of this theorem are discussed at some length. In this section we shall give only a brief description of the method and indicate one simple application to Monte Carlo calculations.

Because the method is difficult to describe directly within the context of Monte Carlo applications, we shall describe its application to the estimate of definite integrals. Accordingly, let x be a uniformly distributed random variable on $0 \leq x \leq 1$ and let $\xi(x)$ be any positive, integrable function. If n independent samples ρ_i are drawn from the uniform distribution (using pseudo-random numbers, say) then

$$\bar{\xi}_n = \frac{1}{n} \sum_{i=1}^{n} \xi(\rho_i) \tag{3.10.1}$$

is an unbiased estimator of

$$\mu = \int_0^1 \xi(x)\, dx \tag{3.10.2}$$

with variance

$$V[\bar{\xi}_n] = \frac{1}{n} V[\xi]$$

$$= \frac{1}{n} \left\{ \int_0^1 \xi^2(x)\, dx - \mu^2 \right\}$$

$$= \frac{1}{n} V. \tag{3.10.3}$$

As a special, simple example of the antithetic variates method, consider the estimator

$$\xi_1(\rho) = \frac{\xi(\rho) + \xi(1 - \rho)}{2}. \tag{3.10.4}$$

The variance of ξ_1, $V[\xi_1] = V_1$, will certainly depend on the behavior of the function ξ on $0 \le x \le 1$. However, a simple application of Schwartz's inequality gives

$$V_1 \le V, \tag{3.10.5}$$

since

$$V_1 = \frac{1}{4} \int_0^1 \xi^2(x)\, dx + \frac{1}{4} \int_0^1 \xi^2(1 - x)\, dx$$

$$+ \frac{1}{2} \int_0^1 \xi(x)\xi(1 - x)\, dx - \mu^2 = V/2 + C,$$

where

$$C = \frac{1}{2} \int_0^1 \xi(x)\xi(1 - x)\, dx - \frac{\mu^2}{2}.$$

But

$$C \le \left[\int_0^1 \xi^2(x)\, dx \int_0^1 \xi^2(1 - x)\, dx \right]^{1/2} - \frac{\mu^2}{2} = V/2,$$

which gives (3.10.5).

We note, however, that the evaluation of ξ_1 requires (potentially) twice as much work as the evaluation of ξ. Thus, one really would like to have

$$V_1 \le V/2 \tag{3.10.6}$$

to guarantee a gain in efficiency. Along these lines, we have the following result.

Theorem 3.13 If ξ is a monotone function of x, then $V_1 \leq V/2$.

The proof depends on the following lemma.

Lemma 3.2 Let $\xi(x)$ be a nonnegative, monotone nondecreasing (nonincreasing) function and let $q(x)$ be a monotone nondecreasing (nonincreasing) function with $\int_0^1 q(x)\, dx = 0$. Then $\int_0^1 \xi(x)q(x)\, dx \geq 0$.

To prove the lemma, we choose x_0, $0 \leq x_0 \leq 1$ so that $q(x_0) = 0$ and

$$\int_0^{x_0} q(x)\, dx = -\int_{x_0}^1 q(x)\, dx = a.$$

By the mean value theorem, there is a θ, $0 \leq \theta \leq x_0$, such that

$$\int_0^{x_0} \xi(x)q(x)\, dx = \xi(\theta) \int_0^{x_0} q(x)\, dx,$$

and there is a ϕ, $x_0 \leq \phi \leq 1$, so that

$$\int_{x_0}^1 \xi(x)q(x)\, dx = \xi(\phi) \int_{x_0}^1 q(x)\, dx.$$

Then

$$\int_0^1 \xi(x)q(x)\, dx = \xi(\theta)a - \xi(\phi)a$$
$$= [\xi(\theta) - \xi(\phi)]a.$$

But $\xi(\theta) - \xi(\phi)$ and a have the same sign, which proves the lemma.

The theorem follows easily from the lemma by choosing

$$q(x) = 1 - \frac{\xi(1 - x)}{\int_0^1 \xi(1 - x)\, dx}. \tag{3.10.7}$$

Theorem 3.13 may sometimes be applied to cases where ξ is piecewise monotone to obtain a gain in efficiency.

The formula (3.10.4) is easily modified to yield other estimators. One may use

$$\bar{\xi}(\rho) = \alpha\xi(\alpha\rho) + (1 - \alpha)\xi[1 - (1 - \alpha)\rho], \tag{3.10.8}$$

which is particularly effective when ξ is monotone, or, as suggested by Goertzel and Kalos (Ref. 21), the estimator

$$\bar{\xi}(\rho) = \alpha\xi(\alpha\rho) + (1 - \alpha)\xi[\alpha + (1 - \alpha)\rho], \tag{3.10.9}$$

which may be effective when ξ is not necessarily monotone.

An application to Monte Carlo estimates of resonance integrals is along the following lines. The number ρ is a pseudo-random number which is used to determine a starting energy E. The random variable $\xi(\rho)$ is essentially the probability that a neutron, born at energy E, will be captured in the resonances.

Here, of course, ξ may be a very complicated function of the variable ρ, but in an analog calculation of resonance integrals one would expect ξ to be roughly a monotone nondecreasing function of E. That is, neutrons born at higher energies tend to have larger expected contributions to the resonance integral than those born at low energies, because they "see" more energies at which resonances can occur. Even in the superposition calculation of resonance integrals detailed in Chapter 6, this tends to be so although the source distribution itself is quite different from the physical one there. Then, if one uses a pseudo-random number ρ to determine an energy E and the number $(1 - \rho)$ to determine a second starting energy E', when ρ is close to 0 or 1, E and E' will have a tendency to be at opposite ends of the energy spectrum of the source. Averaging the contributions of the paired particle histories can result in a very effective variance reduction technique.

3.11 OPTIMUM COMBINATIONS OF VARIABLES

We shall close our chapter on standard variance reduction methods with a brief discussion of the following question. Given two random variables, ξ_1 and ξ_2, each of which provides an unbiased estimate of a number I with respect to a fixed random walk process, how may a number c be chosen so that the linear combination

$$\xi = c\xi_1 + (1 - c)\xi_2 \tag{3.11.1}$$

has least variance? If c is a constant, then $E[\xi] = I$ and

$$\begin{aligned} V[\xi] &= c^2 V[\xi_1] + (1 - c)^2 V[\xi_2] + 2c(1 - c) \operatorname{cov}[\xi_1, \xi_2] \\ &\equiv c^2 V_1 + (1 - c)^2 V_2 + 2c(1 - c)V_{12}, \end{aligned} \tag{3.11.2}$$

where $V_1 = V[\xi_1]$, $V_2 = V[\xi_3]$, $V_{12} = \operatorname{cov}[\xi_1, \xi_2]$. Then one may differentiate with respect to c and, setting the derivative to zero, find that

$$c = \frac{V_1 - V_{12}}{V_1 + V_2 - 2V_{12}} \tag{3.11.3}$$

gives a minimum.

Unfortunately, the quantities V_1, V_2, V_{12} are themselves never known in practice but must be estimated by sample moments. This renders the problem very much more complicated, since (3.11.1) becomes a sum of products of random variables.

Halperin (Ref. 34) has studied this question and given a solution based on the assumptions that ξ_1 and ξ_2 are normal and that sample moments are used in place of the unknown population moments. We shall not go through Halperin's analysis as it would take us too far afield. The importance of Halperin's contribution is that he is able to derive exact confidence intervals, based on Student's t-distribution, for the resultant linear combination. By a slightly different analysis he can also derive confidence intervals based on a distribution closely related to the t-distribution. The question as to which system of intervals is preferable remains an open one.

As we have pointed out repeatedly, a variety of estimators of a more or less standard nature may be used in transport calculations. The ones we have developed at length in Chapter 2 which are of primary importance are the capture estimator, the collision estimator, and the track length estimator and its variants. Broadly speaking, for large and heavily absorbing regions the capture estimator has relatively low variance, while for optically thin or weakly absorbing regions collision or track length estimators become superior. It is thus desirable, where storage and other considerations permit, to combine these two types of estimators in some minimum variance fashion. It is our experience that Halperin's method works well in a large variety of cases.

MacMillan (Ref. 35) has studied the performance of combined estimators by analyzing the variances of several estimators and combinations for a special set of model problems. Other experimentation of a similar nature but for more practical problems encountered in nuclear design, is reported in Ref. 36. These results all indicate that it is always profitable to combine estimators (as the theory would lead one to hope) of the two types mentioned above—capture and track length. Exactly which type of track length estimator should be used for a given problem depends not only on the problem but on the specific computer used.

REFERENCES

1. J. M. HAMMERSLEY and D. C. HANDSCOMB, *Monte Carlo Methods*, Methuen and Co. Ltd., London (1964).

2. H. F. TROTTER and J. W. TUKEY, "Conditional Monte Carlo for Normal Samples," *Symposium on Monte Carlo Methods*, ed. H. A. Meyer, John Wiley and Sons, Inc., New York (1956), pp. 64–79.

3. D. W. DRAWBAUGH, "On the Solution of Transport Problems by Conditional Monte Carlo," *Nucl. Sci. Eng.*, **9**, 185 (1961).

4. J. M. HAMMERSLEY, "Conditional Monte Carlo," *J. Assoc. Comp. Mach.*, **3**, 73 (1956).

5. J. G. WENDEL, "Groups and Conditional Monte Carlo," *Ann. Math. Stat.*, **28**, 1048 (1957).

6. S. M. ERMAKOV and V. G. ZOLOTUKHIN, "Polynomial Approximations and the Monte Carlo Method," *Teor. Veroyatnost, Primenen*, **5**, 473 (1960); trans. as *Theor. Prob. Appl.*, **5**, 428–431.

7. D. C. HANDSCOMB, "Remarks on a Monte Carlo Integration Method," *Numerische Math.*, **6**, 261 (1964).

8. M. LEIMDÖRFER, "On the Transformation of the Transport Equation for Solving Deep Penetration Problems by the Monte Carlo Method," *Trans. Chalmers Univ. Technol., Gothenberg,* **286** (1964).

9. F. H. CLARK, "The Exponential Transform as an Importance-Sampling Device—A Review," *ORNL-RSIC*-14, (Jan., 1966).

10. M. LEIMDÖRFER, "A Monte Carlo Method for the Analysis of Gamma Radiation Transport from Distributed Sources in Laminated Shields," *Nukleonik,* **6,** 58 (1964).

11. J. SPANIER, "Monte Carlo Methods and Their Application to Neutron Transport Problems," *Bettis Atomic Power Laboratory Report, WAPD*-195 (July, 1959).

12. G. E. ALBERT, "A General Theory of Stochastic Estimates of the Neumann Series for the Solutions of Certain Fredholm Integral Equations and Related Series," *ORNL*-1508 (1953).

13. W. FELLER, *An Introduction to Probability Theory and Its Applications,* Vol. I, John Wiley and Sons, Inc., New York (1950).

14. C. T. I. TULCEA, "Mesures dans les Espaces Produits," *Atti. Accad. Naz. Lincei Rend., Cl. Sci. Fis. Mat. Nat.,* (8), **7,** 208 (1949).

15. J. SPANIER, "Some Results on Transport Theory and Their Application to Monte Carlo Methods," *J. Math. Anal. Appl.,* **17,** 549 (1967).

16. H. CRAMÉR, *The Elements of Probability Theory,* John Wiley and Sons, Inc., New York (1955).

17. M. H. KALOS, "On the Estimation of Flux at a Point by Monte Carlo," *Nucl. Sci. Eng.,* **16,** 111 (1963).

18. G. GOERTZEL, "Quota Sampling and Importance Functions in Stochastic Solution of Particle Problems," *ORNL*-434 (1949).

19. HERMAN KAHN, *Applications of Monte Carlo,* RAND Corp., AECU-3259 (April, 1954; Revised April, 1956).

20. H. KAHN, "Use of Different Monte Carlo Sampling Techniques," *Symposium on Monte Carlo Methods,* ed. H. A. Meyer, John Wiley and Sons, Inc., New York (1956), pp. 146–190.

21. G. GOERTZEL and M. H. KALOS, "Monte Carlo Methods in Transport Problems," *Progress in Nuclear Energy,* Vol. II, Series I, Physics and Mathematics, ed. D. J. Hughes, J. E. Sanders and J. Horwitz, Pergamon Press, New York (1958), pp. 315–369.

22. M. H. KALOS, "Importance Sampling in Monte Carlo Shielding Calculations," *Nucl. Sci. Eng.,* **16,** 227 (1963).

23. G. E. ALBERT, "A General Theory of Stochastic Estimates of the Neumann Series for the Solutions of Certain Fredholm Integral Equations and Related Series," *ORNL*-1508 (1953).

24. R. A. LIEDTKE and H. STEINBERG, "A Monte Carlo Code for Gamma-Ray Transmission Through Laminated Slab Shields," *WADC-TR* 58–80 (1958).

25. M. LEIMDÖRFER, "The Backscattering of Fast Neutrons from Plane and Spherical Reflectors," *Trans. Chalmers Univ. Technol., Gothenberg,* **288** (1964).

26. M. LEIMDÖRFER, "On the Use of Monte Carlo Methods of Calculating the Deep Penetration of Neutrons in Shields," *Trans. Chalmers Univ. Technol., Gothenberg*, **287** (1964).

27. L. A. BEACH and R. H. THEUS, "Stochastic Calculations of Gamma Ray Diffusion," *Symposium on Monte Carlo Methods*, ed. H. A. Meyer, John Wiley and Sons, Inc., New York (1956), pp. 103–122.

28. N. J. CURLEE, JR. and L. A. ONDIS II, "The Use of Correlated Sampling in Monte Carlo Calculation of Changes in Thermal Absorption," *Trans. Am. Nucl. Soc.*, **7**, 2 (1964).

29. H. STEINBERG and R. ARONSON, "Monte Carlo Calculations of Gamma Ray Penetration," *Technical Research Group, Inc.*, *WADC-TR-59-771* (Aug., 1960).

30. J. F. OLHOEFT, "The Doppler Effect for a Non-uniform Temperature Distribution in Reactor Fuel Elements," Ph.D. Dissertation, Univ. Michigan (1963).

31. J. M. HAMMERSLEY and K. W. MORTON, "A New Monte Carlo Technique: Antithetic Variates," *Proc. Camb. Phil. Soc.*, **52**, 449 (1956).

32. J. M. HAMMERSLEY and J. G. MAULDON, "General Principles of Antithetic Variates," *Proc. Camb. Phil. Soc.*, **52**, 476 (1956).

33. D. C. HANDSCOMB, "Proof of the Antithetic Variates Theorem for n > 2," *Proc. Camb. Phil. Soc.*, **54**, 300 (1958).

34. M. HALPERIN, "Almost Linearly-Optimum Combination of Unbiased Estimates," *J. Am. Stat. Assoc.*, **56**, 36 (1961).

35. D. B. MACMILLAN, "Comparison of Statistical Estimators for Neutron Monte Carlo Calculations," *Nucl. Sci. Eng.*, **26**, 366 (1966).

36. E. M. GELBARD, L. A. ONDIS II, and J. SPANIER, "A New Class of Monte Carlo Estimators," *J. SIAM Appl. Math.*, **14**, 697 (1966).

4

Use of Superposition and Reciprocity In One-Energy Problems

4.1 INTRODUCTION

We hope that, at this point, the reader understands the most basic principles of Monte Carlo and has at his disposal a repertoire of conventional Monte Carlo techniques. It is possible, however, that he does not yet appreciate the power or the versatility of Monte Carlo. To bring out more clearly the capabilities of Monte Carlo, as well as its weaknesses, we must put the method to some use. In preceding chapters we have discussed Monte Carlo problems in rather general terms. Now we feel we must take a more practical course. Therefore, we shall fix our attention on specific neutron transport problems, simple problems initially, of various types. In the work which follows we can rely on much of the theory discussed earlier, but it will be necessary also to introduce new methods as we encounter new difficulties. We have reached a stage, then, where our outlook and our subject matter have both been slightly altered. Having (more or less) completed our analysis of the general principles of Monte Carlo, we find that in the future we shall have to deal with specialized techniques designed to solve special problems. As our subject matter changes, we feel it is appropriate to make corresponding changes in the style of our exposition. The reader will note, for example, that in Chapters 1 through 3 we based our arguments almost exclusively on the integral form of the transport equations. So long as we were primarily concerned with the fundamental theory underlying Monte Carlo methods this form seemed most convenient. Now, however, we shift our point of view. The transport equation itself will come to play a new large role in our considerations and it will be necessary, frequently, to put this equation through complicated manipulations. We shall also need to develop for our use some simple (though valuable) rudiments of transport theory. For the task ahead of us it is customary, and convenient too, to work with the integro-differential transport equation, as we shall do.

Throughout the final chapters of this book, we have two main objectives. First, of course, we are interested in certain special problems for their own sake: we want to show how these problems can be solved efficiently. Still we have another, broader, goal. We wish to demonstrate, and to stress, that Monte Carlo need not be used blindly, but that Monte Carlo methods can often be tailored precisely to our needs.

The first problem we shall deal with is the simplest, though perhaps the most instructive. We consider a cell of any sort (Fig. 4.1)† divided into two regions. Region I contains an isotropic neutron source of uniform density Q_I, while region II is source free:

$$Q(\mathbf{r}, \boldsymbol{\omega}) = \begin{cases} \dfrac{1}{4\pi} Q_I, & \mathbf{r} \in R_I, \\[2mm] 0\ , & \mathbf{r} \in R_{II}. \end{cases}$$

It is assumed here, and throughout Chapter 4, that all neutrons move with the same speed. We are asked to compute the average flux in each region, or alternatively, the region I and region II absorption rates.

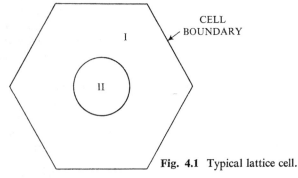

CELL BOUNDARY

I

II

Fig. 4.1 Typical lattice cell.

In principle, of course, such a problem can always be solved conventionally. One might, for example, generate sample histories by analog methods: combined absorption and track length estimators could be used to compute the flux averages. At times this most straightforward approach is as good as any other, but one can easily construct problems for which it is ill-suited. Suppose, for example, that $\Sigma_{aI} \neq 0$. Then, if region II is sufficiently small, it is improbable that any given starter will enter it. In such circumstances an analog computation of $\bar{\phi}_{II}$, the average flux in region II, via any simple estimator may be costly, even on a fast computer. It is characteristic

---

† A cell is a structural unit of a lattice. The lattice is composed of identical cells adjoining each other, and arranged in some regular, space-filling, pattern. In Fig. 4.1 we have depicted a hexagonal cell containing a cylindrical fuel rod. The rod is shown in cross section.

of Monte Carlo that a method devised to treat a broad class of problems is inefficient in selected special cases. It is also typical that, when details of a problem are specified, special Monte Carlo techniques can be used to solve it. How, then, can one compute $\bar{\phi}_{\text{II}}$ given that region II is small? We have found, in practice, that two methods are equally feasible. One, the adjoint method, has already been mentioned in Chapters 2 and 3. Here we shall describe a second method, namely, the superposition method. These two methods will be compared in the last section of this chapter.

4.2 THE SUPERPOSITION METHOD

The superposition method is based on the superposition principle, which in transport theory takes the following form. *The flux produced by the sum of two sources is the sum of the fluxes produced, separately, by each source.* Thus, if the flux F_1 satisfies the equation

$$\boldsymbol{\omega} \cdot \boldsymbol{\nabla}F_1(E, \mathbf{r}, \boldsymbol{\omega}) + \Sigma_T(E, \mathbf{r})F_1(E, \mathbf{r}, \boldsymbol{\omega})$$
$$= \int \Sigma_s(E' \to E, \mathbf{r}, \boldsymbol{\omega} \cdot \boldsymbol{\omega}')F_1(E', \mathbf{r}, \boldsymbol{\omega}')\, dE'\, d\boldsymbol{\omega}' + Q_1(E, \mathbf{r}, \boldsymbol{\omega})$$

and, correspondingly, if

$$\boldsymbol{\omega} \cdot \boldsymbol{\nabla}F_2(E, \mathbf{r}, \boldsymbol{\omega}) + \Sigma_T(E, \mathbf{r})F_2(E, \mathbf{r}, \boldsymbol{\omega})$$
$$= \int \Sigma_s(E' \to E, \mathbf{r}, \boldsymbol{\omega} \cdot \boldsymbol{\omega}')F_2(E', \mathbf{r}, \boldsymbol{\omega}')\, dE'\, d\boldsymbol{\omega}' + Q_2(E, \mathbf{r}, \boldsymbol{\omega}),$$

then

$$\boldsymbol{\omega} \cdot \boldsymbol{\nabla}F_T(E, \mathbf{r}, \boldsymbol{\omega}) + \Sigma_T(E, \mathbf{r})F_T(E, \mathbf{r}, \boldsymbol{\omega})$$
$$= \int \Sigma_s(E' \to E, \mathbf{r}, \boldsymbol{\omega} \cdot \boldsymbol{\omega}')F_T(E', \mathbf{r}, \boldsymbol{\omega}')\, dE'\, d\boldsymbol{\omega}' + Q_1(E, \mathbf{r}, \boldsymbol{\omega}) + Q_2(E, \mathbf{r}, \boldsymbol{\omega}).$$

Here

$$F_T(E, \mathbf{r}, \boldsymbol{\omega}) = F_1(E, \mathbf{r}, \boldsymbol{\omega}) + F_2(E, \mathbf{r}, \boldsymbol{\omega})$$

and, in the notation of earlier chapters,

$$\Sigma_s(E' \to E, \mathbf{r}, \boldsymbol{\omega} \cdot \boldsymbol{\omega}') \equiv \Sigma_t(\mathbf{r}, E')C(\mathbf{E}', \mathbf{E}; \mathbf{r}),$$

where $\mathbf{E} = E\boldsymbol{\omega}$, $\mathbf{E}' = E'\boldsymbol{\omega}'$. We shall invoke the superposition principle repeatedly to transform a difficult Monte Carlo problem into one which is more tractable. In the original and transformed problems all cross sections are identical: only the source densities will be different. Henceforth the transformed problem will be called "problem β"; we designate the original problem as "problem α." By way of illustration we now apply the superposition method to our two-region cell problem.

In problem α there is a source, Q_I, in region I, but no source in region II:

$$Q_\alpha(\mathbf{r}, \omega) = \begin{cases} \dfrac{1}{4\pi} Q_\mathrm{I}, & \mathbf{r} \in R_\mathrm{I}, \\[2mm] 0, & \mathbf{r} \in R_\mathrm{II}. \end{cases}$$

In problem β, on the other hand, we postulate a source

$$Q_\beta(\mathbf{r}, \omega) = \begin{cases} 0, & \mathbf{r} \in R_\mathrm{I}, \\[2mm] \dfrac{1}{4\pi} Q_\mathrm{II}, & \mathbf{r} \in R_\mathrm{II}. \end{cases}$$

We now suppose that

$$Q_\mathrm{II} = \Sigma_{a\mathrm{II}} Q_\mathrm{I} / \Sigma_{a\mathrm{I}}, \tag{4.2.1}$$

and define a quantity $F_T(\mathbf{r}, \omega)$:

$$F_T(\mathbf{r}, \omega) = F_\alpha(\mathbf{r}, \omega) + F_\beta(\mathbf{r}, \omega). \tag{4.2.2}$$

Now, $F_\alpha(\mathbf{r}, \omega)$ and $F_\beta(\mathbf{r}, \omega)$ are, respectively, the angular fluxes in problems α and β. By superposition, then, $F_T(\mathbf{r}, \omega)$ is the flux in a cell containing both Q_I and Q_II. It follows from (4.2.1) that in this cell the ratio of source strength to absorption cross section is independent of position. Consequently (as can be shown by direct substitution into the transport equation),

$$F_T(\mathbf{r}, \omega) = \frac{1}{4\pi} \frac{Q_\mathrm{I}}{\Sigma_{a\mathrm{I}}} = \frac{1}{4\pi} \frac{Q_\mathrm{II}}{\Sigma_{a\mathrm{II}}} = \frac{1}{4\pi} \phi_T, \tag{4.2.3}$$

where ϕ_T is a position independent scalar flux. From Eq. (4.2.3) we conclude that

$$F_\alpha(\mathbf{r}, \omega) = \frac{1}{4\pi} \frac{Q_\mathrm{II}}{\Sigma_{a\mathrm{II}}} - F_\beta(\mathbf{r}, \omega), \tag{4.2.4}$$

and

$$\int_\mathrm{II} \Sigma_{a\mathrm{II}} \phi_\alpha(\mathbf{r})\, d\mathbf{r} = Q_\mathrm{II} V_\mathrm{II} - \int_\mathrm{II} \Sigma_{a\mathrm{II}} \phi_\beta(\mathbf{r})\, d\mathbf{r} = \int_\mathrm{I} \Sigma_{a\mathrm{I}} \phi_\beta(\mathbf{r})\, d\mathbf{r}. \tag{4.2.5}$$

Here, of course,

$$\phi_\alpha(\mathbf{r}) \equiv \int F_\alpha(\mathbf{r}, \omega)\, d\omega, \qquad \phi_\beta(\mathbf{r}) \equiv \int F_\beta(\mathbf{r}, \omega)\, d\omega.$$

Equation (4.2.4) relates the fluxes in problems α and β. Equation (4.2.5) implies that the problem α absorption rate in region II is equal to the problem β absorption rate in I. Hence, we solve problem α completely in solving problem β.

Upon the introduction of transfer probabilities, Eq. (4.2.5) takes on a more familiar form. Let

$$P_{\text{I}\to\text{II}} \equiv \int_{\text{II}} \Sigma_{a\text{II}}\phi_\alpha(\mathbf{r})\, d\mathbf{r}/Q_{\text{I}}V_{\text{I}},$$

$$P_{\text{II}\to\text{I}} \equiv \int_{\text{I}} \Sigma_{a\text{I}}\phi_\beta(\mathbf{r})\, d\mathbf{r}/Q_{\text{II}}V_{\text{II}}.$$

Note that $P_{\text{I}\to\text{II}}$ is the transfer probability from I to II, i.e. the probability that a particle born in region I will be captured in region II. Similarly, $P_{\text{I}\to\text{II}}$ is the transfer probability from II to I. Making use of Eq. (4.2.1), we find that

$$V_{\text{I}}\Sigma_{a\text{I}}P_{\text{I}\to\text{II}} = V_{\text{II}}\Sigma_{a\text{II}}P_{\text{II}\to\text{I}}. \tag{4.2.6}$$

Equation (4.2.6) can be (and usually is) derived from the reciprocity relation (see Section 4.10), but Eq. (4.2.4), based on superposition, contains more information than Eq. (4.2.6). Through the use of Eq. (4.2.4) one may deduce the spatial distribution of absorptions in problem α from the flux in problem β. Use of the reciprocity relation, as we shall see later, yields only the net absorption rates in each region.

At this point it is natural to ask: "When should one solve problem α directly, and when should one treat problem β instead?" It is impossible to answer this question precisely, for two reasons. First, there is no simple way to determine the variance as a function of the number of histories in either problem. The variance of the absorption estimator is amenable to straightforward analysis because of its binomial nature, but the variance of more complicated estimators is not. Second, in the case at hand, there is no simple way to estimate the average computing time per history, and this time may be quite different in the original and transformed problems.

However, in order to gain some insight into the capabilities of the superposition method, we shall overlook these difficulties. We postulate that only the absorption estimator is to be used in both the original and transformed problems, and we assume that the computing time per history is the same in both problems. Since the absorption estimator is binomial,

$$\sigma_\alpha^2 = \frac{P_{\text{I}\to\text{II}}(1 - P_{\text{I}\to\text{II}})}{N_\alpha} \tag{4.2.7}$$

and

$$\sigma_\beta^2 = \frac{P_{\text{II}\to\text{I}}(1 - P_{\text{II}\to\text{I}})}{N_\beta}. \tag{4.2.8}$$

Here σ_α^2 and σ_β^2 are variances in the original and transformed problems, respectively. Similarly, N_α and N_β are numbers of starters. If the relative errors in problems α and β are equal, then

$$\frac{\sigma_\alpha}{P_{\text{I}\to\text{II}}} = \frac{\sigma_\beta}{P_{\text{II}\to\text{I}}},$$

so that

$$\frac{N_\alpha}{N_\beta} = \frac{P_{II \to I}(1 - P_{I \to II})}{P_{I \to II}(1 - P_{II \to I})}. \tag{4.2.9}$$

According to our assumptions, the ratio N_α/N_β is also the ratio of running times in problems α and β. Thus, it is advantageous to solve problem β instead of problem α if

$$P_{II \to I} > P_{I \to II}, \tag{4.2.10}$$

or, from Eq. (4.2.6), if

$$V_I \Sigma_{aI} > V_{II} \Sigma_{aII}. \tag{4.2.11}$$

In view of the crudeness of our assumptions, we cannot expect that Eq. (4.2.11) will tell us unerringly which Monte Carlo method to use. It is no more than a rule of thumb which *roughly* delineates the class of problems for which superposition should be useful.

4.3 THE SURFACE SOURCE METHOD

Equation (4.2.11) indicates the range of applicability of the superposition method *in its simplest form*. However, the method can be refined so as to extend this range. In important special cases the effectiveness of superposition as a variance reducing device can be greatly enhanced. We show, first, how this can be done when Σ_{sII}, the scattering cross sections in region II, is equal to zero, and, in later sections, when $\Sigma_{sII} \neq 0$.

It will be seen from Eq. (4.2.11) that the use of the superposition method, *as it is formulated in Section* 4.2, becomes unprofitable when Σ_{aII} is very large. If Σ_{aII} is sufficiently large, few starters born in II will emerge to give us information on the leakage rate. Fortunately, it is easy to modify the Monte Carlo process so that more starters will leak. One might, for example, force each starter to escape from the fuel uncollided. Each starter's weight would then be adjusted appropriately.† Here, however, a somewhat different method seems more effective. This method (which we shall call the "surface source" method) is outlined below.

The idea underlying the surface source method is quite simple. We replace the volume source within region II with a source distributed over its surface. The surface and volume sources are so related as to create the same flux in region I.

We must show now that it is possible to construct a surface source which is equivalent to the given volume source. To this end we proceed as follows. We fix our attention, in problem β, on the flux, $F_u(\mathbf{r}, \mathbf{\omega})$, generated by "uncollided unreturned neutrons." Uncollided unreturned neutrons are

---

† It should be clear that the starting weight will be equal to $e^{-\Sigma_{aII}l}$. Here l is the distance, along $\mathbf{\omega}$, between the particle's point of birth and the boundary of II.

those which have never suffered a collision and have never left the fuel. Clearly,

$$\boldsymbol{\omega} \cdot \nabla F_u(\mathbf{r}, \boldsymbol{\omega}) + \Sigma_{aII} F_u(\mathbf{r}, \boldsymbol{\omega}) = \frac{1}{4\pi} Q_{II}, \qquad \mathbf{r} \in R_{II}. \qquad (4.3.1)$$

Suppose that P is a point on the boundary of II, as in Fig. 4.2, and that $\mathbf{n}$ is the outward unit normal at P. Let $\mathbf{r}_p$ be the radius vector to P. Then†

$$F_u(\mathbf{r}_p, \boldsymbol{\omega}) \equiv F_p(\boldsymbol{\omega}) = 0, \qquad \boldsymbol{\omega} \cdot \mathbf{n} \equiv \mu \leq 0. \qquad (4.3.2)$$

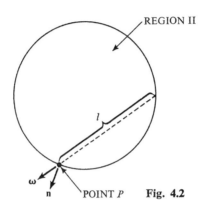

REGION II

POINT P Fig. 4.2

It follows from the integral form of the transport equation (see Eqs. 2.4.20 and 2.4.22) that‡

$$F_p(\boldsymbol{\omega}) = \frac{1}{4\pi} Q_{II} \int_0^l e^{-\Sigma_{aII} l} \, dt$$

$$= \frac{1}{4\pi} Q_{II} (1 - e^{-\Sigma_{aII} l}) / \Sigma_{aII}, \qquad \mu > 0.$$

Having computed the flux due to the volume source we now replace the volume source by a surface source. The density per unit area of the surface source is given by the expression

$$Q(\mathbf{r}_p, \boldsymbol{\omega}) = \begin{cases} \dfrac{\mu}{4\pi} Q_{II} (1 - e^{-\Sigma_{aII} l}) / \Sigma_{aII}, & \mu > 0, \\[2mm] 0, & \mu \leq 0. \end{cases} \qquad (4.3.3)$$

---

† Strictly speaking, the flux $F_u(\mathbf{r}, \boldsymbol{\omega})$ is not yet defined at $\mathbf{r}_p$, but we may write $F_u(\mathbf{r}_p, \boldsymbol{\omega}) \equiv \lim_{\mathbf{r} \to \mathbf{r}_p} F_u(\mathbf{r}, \boldsymbol{\omega}), \mathbf{r} \in R_{II}$.

‡ To derive the expression given here we set $C(E', E; \mathbf{r}) = 0$ in Eq. (2.4.20), so that $\chi(\mathbf{r}, E) = Q(\mathbf{r}, E) = Q_{II}/4\pi$. Our expression for $F_p(\boldsymbol{\omega})$ then follows from (2.4.22).

Again from the integral form of the transport equation, we find that the flux produced by this surface source is given by the expression

$$F_p(\omega) = \frac{1}{4\pi} Q_{\mathrm{II}}(1 - e^{-\Sigma_{a\mathrm{II}}l})/\Sigma_{a\mathrm{II}}, \qquad \mu > 0. \qquad (4.3.4)$$

Thus, we see that the surface and volume sources produce the same exiting flux on the boundary of the absorbing lump.† Therefore, both produce the same flux, and the same absorption rate, in region I.

The reader will note that

$$Q(\mathbf{r}_p, \omega) = \mu F_p(\omega). \qquad (4.3.5)$$

Equation (4.3.5) has a simple physical interpretation. In the presence of a flux $F_p(\omega)$ neutrons with directions ω in $d\omega$ emerge from the fuel at the rate $\mu F_p(\omega)\, d\omega$. The source $Q(\mathbf{r}_p, \omega)$ feeds neutrons into $d\omega$ at precisely the same rate. It is for this reason that the surface and volume sources are equivalent.

All the techniques needed to simulate the surface source distribution have already been developed. We may write

$$Q(\mathbf{r}_p, \omega) = W[Q(\mathbf{r}_p, \omega)/W] \equiv WP(\mathbf{r}_p, \omega),$$

$$W = \int_\tau d\mathbf{r}'_p \int d\omega Q(\mathbf{r}'_p, \omega),$$

where τ is the surface of region II.‡ The function $P(\mathbf{r}_p, \omega)$ can be regarded as a density function for sample starting points, $\mathbf{r}_p$, and direction vectors, ω. If $\mathbf{r}_p$ and ω are drawn from this distribution, each starter will carry the weight W. Otherwise the starting weight must be modified appropriately. Suppose, for example, that starting points are taken from a distribution P' uniform over τ, while outgoing ω's are selected from an isotropic distribution. If A_{II} is the surface area of the fuel,

$$P'(\mathbf{r}_p, \omega) = 1/2\pi A_{\mathrm{II}}.$$

Each starter must then be assigned the weight

$$W(\mathbf{r}_p, \omega) = WP(\mathbf{r}_p, \omega)/P'(\mathbf{r}_p, \omega) \equiv Wf'(\mathbf{r}_p, \omega),$$

$$W(\mathbf{r}_p, \omega) = Q_{\mathrm{II}}A_{\mathrm{II}}\mu[1 - e^{-\Sigma_{a\mathrm{II}}l}]/2\Sigma_{a\mathrm{II}}. \qquad (4.3.6)$$

Thus, the sample weights will fluctuate from history to history.

---

† The use of a fictitious surface source to produce a specified surface flux is discussed in Ref. 1.

‡ If region II has an infinite surface area some of our arguments will have to be changed slightly. For example, if the fuel region is a cylinder, τ is the perimeter of its cross section in a plane normal to its axis.

Now, it has been pointed out in Chapter 2 that fluctuations in the neutron weights will often raise the variance in Monte Carlo estimates. Usually it is advisable to eliminate such fluctuations and this can sometimes be done through rejection.† Suppose, for example, that $f'(\mathbf{r}_p, \boldsymbol{\omega})$ has an upper bound. Then we may proceed as follows. We select a starting point and a starting $\boldsymbol{\omega}$ from $P'(\mathbf{r}_p, \boldsymbol{\omega})$ and compute $f'(\mathbf{r}_p, \boldsymbol{\omega})$. Let $\hat{f}'$ be the least upper bound of f' and compute the ratio $R \equiv f'(\mathbf{r}_p, \boldsymbol{\omega})/\hat{f}'$. Pick a random number, ρ. If $\rho < R$, accept the starter; otherwise reject it. All surviving starters carry the weight W. Rejected starters do not contribute, in any way, to flux or activation estimates. They are simply ignored. We see that

$$RA \equiv \int_{\mathrm{I}} \Sigma_{a\mathrm{II}}\phi_\alpha(\mathbf{r}) \, d\mathbf{r} = \int_{\mathrm{II}} \Sigma_{a\mathrm{I}}\phi_\beta(\mathbf{r}) \, d\mathbf{r} = W - I, \qquad (4.3.7)$$

where I is equal to W times the average probability that a starter will be absorbed in the fuel.

In the foregoing discussion we have tacitly assumed that the weight, W, is known. But we have not yet indicated how W is to be computed. It is clear that

$$W \equiv \int_\tau d\mathbf{r}'_p \int d\boldsymbol{\omega} Q(\mathbf{r}'_p, \boldsymbol{\omega}) = \int_\tau d\mathbf{r}'_p \int d\boldsymbol{\omega}\mu F_p(\boldsymbol{\omega})$$

$$= J.$$

Here J is precisely the number of neutrons which would leak from the fuel if it were surrounded by a perfectly black absorber.‡ If $P_{0(\mathrm{II})}$ is the escape probability from the fuel,§ then

$$W = J = Q_{\mathrm{II}}V_{\mathrm{II}}P_{0(\mathrm{II})}.$$

Thus, we can determine W if we know $P_{0(\mathrm{II})}$.

Escape probabilities from absorbing slabs, spheres, hemispheres, and infinite cylinders have been tabulated by Case, de Hoffmann, and Placzek (Ref. 1). Obviously, if we can read $P_{0(\mathrm{II})}$ from tables the calculation of W is trivial. Otherwise it is necessary to compute $P_{0(\mathrm{II})}$ and this can be done in many ways (see Ref. 1, for example). In the next section we shall describe a Monte Carlo procedure which we have often used for the computation of escape probabilities.

From arguments similar to those used in Chapter 1 (see pp. 26, 27), we know that E, the efficiency of our rejection process, is equal to $\hat{f}'/\hat{f}'$. Here $\hat{f}'$

---

† See p. 26.
‡ A "black absorber" is a medium which absorbs all neutrons which enter it.
§ The escape probability from a region is the average probability that a neutron drawn from a uniform source in the region will escape into a black absorber surrounding this region.

is the average of f' over all $\mathbf{r}_p$ and $\boldsymbol{\omega}$. Since

$$f'(\mathbf{r}_p, \boldsymbol{\omega}) \equiv P(\mathbf{r}_p, \boldsymbol{\omega})/P'(\mathbf{r}_p, \boldsymbol{\omega}), \tag{4.3.8}$$

$$\bar{f}' \equiv \int_\tau d\mathbf{r}_p \int d\boldsymbol{\omega} f'(\mathbf{r}_p, \boldsymbol{\omega}) P'(\mathbf{r}_p, \boldsymbol{\omega}) = \int_\tau d\mathbf{r}_p \int d\boldsymbol{\omega} P(\mathbf{r}_p, \boldsymbol{\omega})$$

$$= 1. \tag{4.3.9}$$

From (4.3.8) it follows that

$$f'(\mathbf{r}_p, \boldsymbol{\omega}) = \frac{\mu A_{\mathrm{II}}}{2\Sigma_{a\mathrm{II}} V_{\mathrm{II}} P_{0(\mathrm{II})}} (1 - e^{-\Sigma_{a\mathrm{II}} l}).$$

Therefore,

$$E = \left[\max_{\omega, \mathbf{r}_p} \frac{\mu A_{\mathrm{II}}}{2\Sigma_{a\mathrm{II}} V_{\mathrm{II}} P_{0(\mathrm{II})}} (1 - e^{-\Sigma_{a\mathrm{II}} l}) \right]^{-1}.$$

For

a) a slab of thickness t, or

b) a cylinder of diameter t,

$$\max_{\omega, \mathbf{r}_p} \mu(1 - e^{-\Sigma_{a\mathrm{II}} l}) = 1 - e^{-\Sigma_{a\mathrm{II}} t},$$

so that

$$E = \left[\frac{A_{\mathrm{II}}}{2\Sigma_{a\mathrm{II}} V_{\mathrm{II}} P_{0(\mathrm{II})}} (1 - e^{-\Sigma_{a\mathrm{II}} t}) \right]^{-1}. \tag{4.3.10}$$

With the aid of Eq. (4.3.10) it can be shown that

a) $E \geq 0.500$ for slabs and

b) $E \geq 0.469$ for cylinders.

All in all, what have we gained through the surface source method? If we start neutrons uniformly *throughout* the lump, they escape with probability $P_{0(\mathrm{II})}$; but if, instead, the surface source method is used with rejection roughly half the starters escape. Therefore, if $P_{0(\mathrm{II})} \ll \frac{1}{2}$, the surface source yields much more information per starter than the volume source does. It is precisely for this reason that the surface source method is sometimes very helpful.

Of course, when a surface source method is used, our earlier analysis of the efficiency of superposition is not applicable. However, it is easy to derive a rule of thumb to replace Eq. (4.2.11) when the surface source technique is used and is coupled with rejection. To this end we note, first, that in problem β

$$J_\beta = Q_{\mathrm{II}} V_{\mathrm{II}} P_{\mathrm{II} \rightarrow \mathrm{I}}. \tag{4.3.11}$$

Here J_β is the total number of neutrons absorbed, per second, in region I.

If we write

$$P_{II \to I} = P_{0(II)} \Pi_{II \to I},$$

then $\Pi_{II \to I}$ is the conditional probability† that a neutron, having emerged from region II, will be absorbed in I. Now, making the same assumption as in Section 4.2 (see p. 143), we find that

$$\sigma_\beta^2 = Q_{II}^2 V_{II}^2 P_{0(II)}^2 \Pi_{II \to I}(1 - \Pi_{II \to I}),$$

for, in the surface source method, $\Pi_{II \to I}$ is the probability which the Monte Carlo computation estimates. Further,

$$\frac{\sigma_\beta^2}{J_\beta^2} = \frac{1 - \Pi_{II \to I}}{\Pi_{II \to I}}.$$

Arguing as in Section 4.2,

$$\frac{N_\alpha}{N_\beta} = \frac{\Pi_{II \to I}(1 - P_{I \to II})}{P_{I \to II}(1 - \Pi_{II \to I})},$$

if we do not include, in N_β, the number of rejected starters. We conclude that superposition with a surface source is more efficient than conventional Monte Carlo if

$$\frac{1 - \Pi_{II \to I}}{\Pi_{II \to I}} < \frac{1 - P_{I \to II}}{P_{I \to II}} \tag{4.3.12}$$

or, in other words, if

$$\Pi_{II \to I} > P_{I \to II}. \tag{4.3.13}$$

But $\Pi_{II \to I} = P_{II \to I}/P_{0(II)}$, so that Eq. (4.3.13) can be written in the form

$$\frac{P_{II \to I}}{P_{0(II)}} > P_{I \to II}. \tag{4.3.14}$$

Making use of the reciprocity relation (Eq. 4.2.6), we are led to the condition

$$\frac{V_I \Sigma_{aI}}{P_{0(II)}} > V_{II} \Sigma_{aII}. \tag{4.3.15}$$

If we include the number of rejected starters in N_β, we derive an inequality which is only slightly different:

$$\frac{E V_I \Sigma_{aI}}{P_{0(II)}} > V_{II} \Sigma_{aII}. \tag{4.3.16}$$

Since E is always about 0.5, it matters little in practice whether we adopt (4.3.15) or (4.3.16) as our rule of thumb. When the surface source method is used with superposition, either (4.3.15) or (4.3.16) takes the place of (4.2.11).

---

† See the definitions of conditional probability in Chapter 1, pp. 8 ff

Without rejection the preceding argument is invalid. If the particle weights are allowed to fluctuate, the distribution of absorptions is no longer binomial and no simple analysis seems possible.

It is important to realize that the surface source density can be simulated in many ways. We have chosen to take $\boldsymbol{\omega}$ from a uniform distribution on a unit hemisphere. We might instead have selected $\mu \equiv \boldsymbol{\omega} \cdot \mathbf{n}$ from a cosine distribution, i.e. from the density function $P(\mu) = 2\mu$. In this case, mimicking Eqs. (4.3.6) and (4.3.8), we find that

$$P''(\mathbf{r}_p, \boldsymbol{\omega}) = 2\mu/2\pi A_{\text{II}}$$

and

$$f'' \equiv P(\mathbf{r}_p, \boldsymbol{\omega})/P''(\mathbf{r}_p, \boldsymbol{\omega}),$$

so that

$$f'' = \left[\frac{F_p(\boldsymbol{\omega})}{2W}\right]\Big/\left[\frac{1}{2\pi A_{\text{II}}}\right] = \frac{A_{\text{II}}}{4\Sigma_{a\text{II}} V_{\text{II}} P_{0(\text{II})}}(1 - e^{-\Sigma_{a\text{II}} l}).$$

It is clear† that

$$\bar{\bar{f}}'' \equiv \int_\tau d\mathbf{r}_p \int d\boldsymbol{\omega} \frac{P(\mathbf{r}_p, \boldsymbol{\omega})}{P''(\mathbf{r}_p, \boldsymbol{\omega})} P''(\mathbf{r}_p, \boldsymbol{\omega})$$

$$= 1. \tag{4.3.17}$$

Having modified the sampling procedure, we find that E, the efficiency of the new rejection process, is now given by the expression

$$E = 4\Sigma_{a\text{II}} V_{\text{II}} P_{0(\text{II})}/A_{\text{II}}.$$

In the next section‡ we show that

$$P_{0(\text{II})} \to A_{\text{II}}/4\Sigma_{a\text{II}} V_{\text{II}} \tag{4.3.18}$$

as $\Sigma_{a\text{II}} \to \infty$. Therefore, $E \to 1$ as $\Sigma_{a\text{II}} \to \infty$. We see also that

$$\lim_{\Sigma_{a\text{II}} \to 0} E = 0,$$

since

$$\lim_{\Sigma_{a\text{II}} \to 0} P_{0(\text{II})} = 1.$$

Thus, this new sampling method is much better than the old for black lumps, but much worse for lumps which are almost transparent, i.e. which absorb very weakly. For practical computation we prefer the first method, since it is fairly good for all lumps. However, to simplify our exposition, we may sometimes work with the second instead.

---

† Throughout this chapter we use bars to indicate averages of different types. To avoid confusion we let a single bar designate averages over $P'(\mathbf{r}_p, \boldsymbol{\omega})$, while a double bar signifies that the average is taken over $P''(\mathbf{r}_p, \boldsymbol{\omega})$.

‡ See discussions leading to Eq. (4.4.6).

The surface source method is a combination of analytic and Monte Carlo procedures. It is a common failing of such semi-analytic methods that, while decreasing the variance for a given number of histories, they increase the running time per history. Characteristically, they require the computation of exponentials or other transcendental functions. This is true of the surface source method, since exponentials occur in the weights.

On any digital computer the calculation of transcendental functions is expensive, and such calculations should be avoided wherever possible. Normally this is done through the use of tables, as discussed in Section 1.5, but it is awkward to store a complete table of the exponential function, particularly when storage space is in short supply. Fortunately, the function which occurs in the weights is not e^{-x}, but $1 - e^{-x}$. This latter function need only be tabulated over a limited range, since we make only a small *relative* error if we set it equal to one for large x. It is still better, however, to tabulate

$$g(x) \equiv x/(1 - e^{-x}), \qquad (4.3.19)$$

a function which is more nearly linear† near the origin than $1 - e^{-x}$. A table giving $g(x)$ at intervals $\Delta x = 0.2$, from $x = 0$ to $x = 7$ contains only 35 values. Yet it is possible, by linear interpolation in such a table, to compute $g(x)$ to within 0.1%. The weight, W, need not be determined more accurately for our purposes. For $x > 7$ one may set $1 - e^{-x}$ equal to one.

4.4 STICKING AND ESCAPE PROBABILITIES

At this point we digress, briefly, from the subject which principally concerns us. We have been dealing with the Monte Carlo treatment of cells containing lumped absorbers. Absorption in lumps may also be treated analytically, as in various theories of resonance capture (Ref. 2) and thermal utilization (Ref. 3). Usually, in such analytic procedures the escape probability plays a central role. Sometimes the sticking probability, G, is introduced instead. In order to make contact with this extensive theoretical domain we shall develop some of the properties of P_0 and G, then show how these quantities may be computed by Monte Carlo.

First we note, from Eq. (4.3.9), that

$$f' \equiv \int_\tau d\mathbf{r}_p \int d\omega f'(\mathbf{r}_p, \omega) P'(\mathbf{r}_p, \omega)$$

$$= \frac{A_{II}}{2\Sigma_{aII} V_{II} P_{0(II)}} \overline{\{\mu(1 - e^{-\Sigma_{aII} l})\}} = 1.$$

---

† Note that $1 - e^{-x} = x + 0(x^2)$. If we neglect the $0(x^2)$-term, the relative error is $0(x)$. On the other hand, $g(x) = 1 + x/2 + 0(x^2)$. If we neglect the $0(x^2)$-term, the relative error in g is $0(x^2)$. This is what we mean when we say that g is "more linear" near the origin than $1 - e^{-x}$.

Therefore,

$$P_{0(II)} = \frac{A_{II}}{2\Sigma_{aII}V_{II}} \overline{\{\mu(1 - e^{-\Sigma_{aII}l})\}}. \tag{4.4.1}$$

Recall that the single bar indicates, here, an average over the surface of the lump, and over an isotropic outgoing $\boldsymbol{\omega}$. We see also, from (4.3.17), that

$$P_{0(II)} = \frac{A_{II}}{4\Sigma_{aII}V_{II}} \overline{\overline{\{(1 - e^{-\Sigma_{aII}l})\}}}. \tag{4.4.2}$$

Suppose that we let Σ_{aII} approach zero. Then

$$e^{-\Sigma_{aII}l} \to 1 - \Sigma_{aII}l$$

and we know that

$$P_{0(II)} \to 1.$$

It follows that†

$$\bar{l}_{II} = 4V_{II}/A_{II}. \tag{4.4.3}$$

The quantity $\bar{l}$ is often referred to as the "mean chord length".

Making use of Eq. (4.4.2), we may write

$$P_{0(II)} = \overline{\overline{\{1 - e^{-\Sigma_{aII}l}\}}}/\Sigma_{aII}\bar{l}_{II} \tag{4.4.4}$$

or, from (4.4.1),

$$P_{0(II)} = \overline{\{2\mu(1 - e^{-\Sigma_{aII}l})\}}/\Sigma_{aII}\bar{l}_{II}. \tag{4.4.5}$$

We may conclude from (4.4.4) or (4.4.5) that

$$P_{0(II)} \to 1/\Sigma_{aII}\bar{l}_{II} \tag{4.4.6}$$

asymptotically as $\Sigma_{aII} \to \infty$. Thus, when an absorbing lump becomes black, its escape probability depends only on its mean chord length and not directly on its shape.

Now assume that the incoming flux at the surface of the absorbing lump is uniform and isotropic. Then the number of absorptions in the lump per entering neutron is, by definition, the sticking probability, G. It is easy to prove that P_0 and G are closely connected. To show how they are related, we postulate that

$$\Sigma_{aI} = \Sigma_{aII} = S_I = S_{II}. \tag{4.4.7}$$

Under these circumstances the flux in the cell is uniform and isotropic. In fact,

$$F(\mathbf{r}, \boldsymbol{\omega}) = 1/4\pi. \tag{4.4.8}$$

---

† We do not claim to have proved Eq. (4.4.3) here. Our argument is a plausibility argument modeled after Dresner's (Ref. 2). The history of Eq. 4.4.3 is discussed in Ref. 4, where the reader will also find a rigorous derivation.

At each point on the lump's surface the entering current is equal to $\frac{1}{4}$:

$$J_{\text{in}} = \int_0^{2\pi} d\phi \int_0^{\pi/2} \left(\frac{1}{4\pi}\right) \cos\theta \sin\theta\, d\theta = \tfrac{1}{4}. \qquad (4.4.9)$$

Therefore, over the whole surface of the lump, the number of neutrons which enter, per second, is equal to $A_{\text{II}}/4$. By definition, $G_{\text{II}}A_{\text{II}}/4$ is the absorption rate in the lump:†

$$G_{\text{II}}A_{\text{II}}/4 = S_{\text{I}}V_{\text{I}}P_{\text{I}\to\text{II}} = \Sigma_{a\text{I}}V_{\text{I}}P_{\text{I}\to\text{II}} = \Sigma_{a\text{II}}V_{\text{II}}P_{\text{II}\to\text{I}}, \qquad (4.4.10)$$

where the V's are volumes.

Suppose that region I is purely absorbing and infinite in all directions. Then $P_{\text{II}\to\text{I}} = P_{0(\text{II})}$. Thus,

$$G_{\text{II}} = \frac{4V_{\text{II}}}{A_{\text{II}}} \Sigma_{a\text{II}}P_{0(\text{II})} = \Sigma_{a\text{II}}P_{0(\text{II})}\bar{l}_{\text{II}}. \qquad (4.4.11)$$

Clearly, the relation between sticking probabilities and escape probabilities involves only the physical properties of the lump itself. Any assumptions we have made regarding region I or Q_{II} are now irrelevant. Upon reviewing our argument, it will be seen that Eq. (4.4.11) is valid for any lump, whether it is a pure absorber or not.

Escape probabilities are used in many branches of reactor physics. Values of P_0 enter into many different computational procedures. We have already pointed out that P_0's have been tabulated for lumps of various one-dimensional shapes. It is sometimes necessary, however, to find escape probabilities for purely absorbing two-dimensional lumps. Such escape probabilities can be computed by Monte Carlo through use of Eq. (4.4.4) or Eq. (4.4.5). We prefer to use Eq. (4.4.5), for reasons which we now discuss.

Suppose we draw $\mathbf{r}_p$ and $\boldsymbol{\omega}$ from $P''(\mathbf{r}_p, \boldsymbol{\omega})$. Then, because of Eq. (4.4.4), the quantity

$$R_1 \equiv (1 - e^{-\Sigma_{a\text{II}}l})/\Sigma_{a\text{II}}\bar{l}_{\text{II}}$$

is certainly an unbiased estimator of $P_{0(\text{II})}$. We could use this estimator, for example, to compute the escape probability for a slab. If the slab is optically thin, then, for most chords,

$$R_1 \doteq \rho_1 \equiv l/\bar{l}_{\text{II}}.$$

But we can show that the variance of ρ_1 is infinite. For

$$\sigma^2(\rho_1) = \overline{(\rho_1^2)} - \overline{(\rho_1)}^2 = \overline{(\rho_1^2)} - 1,$$

---

† The reader will recall that, according to Eq. (4.2.6), $V_{\text{I}}\Sigma_{a\text{I}}P_{\text{I}\to\text{II}} = V_{\text{II}}\Sigma_{a\text{II}}P_{\text{II}\to\text{I}}$.

and, if the thickness of the slab is t,

$$\overline{\overline{(\rho_1^2)}} = \frac{1}{4t^2} \int_0^1 \left(\frac{t^2}{u^2}\right) 2\mu \, d\mu = \infty.$$

We conclude that

$$\lim_{\Sigma_{aII}t \to 0} \sigma^2(R_1) = \infty.$$

On the other hand, if the estimator

$$R_2 \equiv 2\mu(1 - e^{-\Sigma_{aII}l})/\Sigma_{aII}\bar{l}_{II}$$

suggested by Eq. (4.4.5) is used, then

$$\lim_{\Sigma_{aII}t \to 0} \sigma^2(R_2) = 0$$

and

$$\lim_{\Sigma_{aII}t \to \infty} \sigma^2(R_2) = \tfrac{1}{3}.$$

Of course, one would not, generally, use Monte Carlo to compute P_0 for slabs, but we must anticipate that the estimator R_1 will have undesirable features in other geometries as well.†

Note that it is easy to calculate, simultaneously, P_0's for many absorbing lumps of the same size and shape. Let Σ_{aj} be the absorption cross section of the j-th lump. Having drawn r_p and ω from $P'(r_p, \omega)$, we record for each lump the value of R_{2j}, where

$$R_{2j} \equiv 2\mu(1 - e^{-\Sigma_{aj}l})/\Sigma_{aj}\bar{l}.$$

Clearly, $\overline{R_{2j}} = P_{0j}$. In evaluating R_{2j} it is advisable to make use of the g-function tables which we have discussed in Section 4.3.

The method just described may be used to generate short tables of escape probabilities for absorbing lumps of any shape. Fortunately, short tables are adequate for most purposes if in using them one takes advantage of known regularities in the behavior of P_0. We refer, here, to the regularities embodied in the rational approximation (Ref. 2), which we now discuss briefly.

For slabs, cylinders, and spheres,

$$P_0 \doteq 1/(1 + \Sigma_a \bar{l}). \tag{4.4.12}$$

Equation (4.4.12) constitutes the Wigner rational approximation, often involved in the analytic treatment of resonance capture (Ref. 2). This approximation is accurate to within about 20% for one-dimensional lumps. The form of Eq. (4.4.12) suggests that the inverse of P_0 is almost linear in Σ_a, and test calculations have shown that this is true not only for slabs, cylinders, and

---

† The reader will observe that R_1 may be a perfectly good estimator in some cases. This will be true, for example, if the lump is optically thick.

TABLE 4.1
INVERSE OF THE ESCAPE PROBABILITIES
FOR CYLINDERS. THE QUANTITY ρ IS THE
RADIUS OF THE CYLINDER MEASURED IN
MEAN FREE PATHS

ρ	$P_0^{-1}(\rho)$	ρ	$P_0^{-1}(\rho)$
0	1	1.6	3.4952
0.05	1.0652	1.8	3.8591
0.10	1.1299	2.0	4.2292
0.15	1.1950	2.2	4.6045
0.20	1.2610	2.4	4.8450
0.25	1.3278	2.6	5.3668
0.30	1.3957	2.8	5.7524
0.35	1.4646	3.0	6.1402
0.40	1.5346	3.2	6.5295
0.45	1.6058	3.4	6.9204
0.50	1.6780	3.6	7.3126
0.55	1.7513	3.8	3.7059
0.60	1.8256	4.0	8.0998
0.65	1.9011	4.2	8.4947
0.70	1.9775	4.4	8.8896
0.75	2.0548	4.6	9.2851
0.80	2.1334	4.8	9.6815
0.85	2.2125	5.0	10.0776
0.90	2.2930	5.2	10.4745
0.95	2.3739	5.4	10.8719
1.0	2.4561	5.6	11.2689
1.2	2.7920	5.8	11.6673
1.4	1.1389	6.0	12.0642

spheres, but also for wires with rectangular cross sections. Therefore, for lumps of any shape it seems reasonable to tabulate, not P_0, but its inverse. Linear interpolation in a table of P_0^{-1} can be quite accurate even if the tabulation interval is large. By way of illustration we have listed values of P_0^{-1} for cylinders in Table 4.1. Linear interpolation in this 46-entry table yields P_0 to within 0.1%. Beyond the range of the table Eq. (4.4.6) gives the escape probability also to within 0.1%. Such a 46-entry table could be constructed by Monte Carlo very quickly.

4.5 SUPERPOSITION AS A PERTURBATION METHOD

It is often helpful to think of the superposition method, described in Sections 4.2 and 4.3, as a perturbation method.† We imagine that the problem α flux

---

† Here, as in Sections 4.2 and 4.3, we assume that region II is a pure absorber.

impinging on the surface of the lump consists of two components. One is unperturbed by the presence of the lump. This unperturbed flux enters region II and passes through it without hindrance. Inside region II, however, it engenders negative particles which represent absorbed neutrons. Some negative particles, after emerging from the lump, may re-enter it later. These constitute the second component of the impinging flux, the perturbing component. We show now that such a point of view leads directly back to superposition techniques.

If the absorbing material in region II were replaced by a vacuum, the flux in the cell (which would then be equal to the unperturbed flux) would be flat and isotropic. An isotropic incoming flux, $\phi_{in} = Q_I/\Sigma_{aI}$, would exist at the vacuum boundary. We designate the neutrons composing this flux as "unperturbed" neutrons and suppose that unperturbed neutrons are never absorbed in the lump. However, each is accompanied, when it leaves region II, by a particle of weight

$$w = -(1 - e^{-\Sigma_{aII}l}). \qquad (4.5.1)$$

Here l, of course, is the length of the neutron's path within region II.

Since unperturbed neutrons pass through the lump unimpeded, the ingoing and outgoing unperturbed fluxes are equal, i.e.

$$\phi_{out} = \phi_{in} = Q_I/\Sigma_{aI}. \qquad (4.5.2)$$

Therefore, the flux of negatively weighted particles emerging at P and moving along $\boldsymbol{\omega}$ (see Fig. 4.2, p. 145) is given by the expression

$$F_p(\boldsymbol{\omega}) = \frac{1}{4\pi}\frac{Q_I}{\Sigma_{aI}}(1 - e^{-\Sigma_{aII}l}) = \frac{1}{4\pi}\frac{Q_{II}}{\Sigma_{aII}}(1 - e^{-\Sigma_{aII}l}). \qquad (4.5.3)$$

But this is exactly the surface flux (Eq. 4.3.4) which we have simulated in Section 4.3. Thus the surface source method gives us, directly, the perturbing component of the flux.

Negative particles which re-enter the lump *can* be reabsorbed there. Such reabsorptions represent the effect of a flux dip near the lump's surface. The true absorption rate in the lump (the problem α absorption rate) is the net rate at which negative weight leaks out of it, and this in turn is equal to the current out of region II in problem β.

Similar arguments show that the source, Q_{II}, in Section 4.2 may be regarded as a fictitious *volume* source of negative neutrons. These, again, delete the contributions of absorbed neutrons from the unperturbed flux.

When we say that the superposition method is designed to compute perturbations, we do not mean that it involves any approximations. Calculations based on superposition differ from the familiar perturbation calculations in that superposition calculations are exact. Furthermore, the superposition principle may be quite useful even when the perturbation is large: inequality

(4.3.16) may still be satisfied in such a case. Thus, in viewing superposition as a perturbation method, we imply nothing as to its accuracy or effectiveness. On the other hand, this approach does often help us establish (or check) Monte Carlo procedures which embody the superposition principle.

4.6 THE INCLUSION OF SCATTERING

Heretofore we have used surface source simulation only for purely absorbing lumps. Of course, the superposition method can be brought to bear in *any* case if the volume source in region II is sampled directly (see Section 4.2). We are reluctant, however, to forego the demonstrated advantages of surface source simulation. It seems worthwhile, instead, to generalize the surface source procedure, to develop a surface source method applicable when $\Sigma_{sII} \neq 0$. We shall describe two such methods: one will be discussed here and one in Section 4.8 below. There are many other ways to incorporate scattering into surface source calculations. We choose to confine our attention to the two approaches we have found most useful, not only in one-energy problems but also in the computation of resonance escape probabilities. In fact, in all remaining sections of this chapter, and indeed throughout the chapter, the one-energy problem is not our only or our main concern. Our analysis of one-energy problems is a pedagogical device, a device which we have used to introduce, in simple form, the central ideas of Chapters 5 and 6.

First Method. Assume, initially, that the presence of the lump does not perturb the entering flux. Then the entering current is equal to $Q_I A_{II}/4\Sigma_{aI}$ and the collision rate, $(RC)_0$, in the lump is given by the expression†

$$(RC)_0 \equiv \int_{II} \Sigma_{TII} \phi_a(\mathbf{r})\, d\mathbf{r}$$

$$= \frac{1}{4} \frac{Q_I}{\Sigma_{aI}} A_{II} G(\Sigma_{TII}) = \frac{1}{4} \frac{Q_I}{\Sigma_{aI}} A_{II} \Sigma_{TII} \bar{l}_{II} P_0(\Sigma_{TII})$$

$$= \frac{Q_I}{\Sigma_{aI}} \Sigma_{TII} V_{II} P_0(\Sigma_{TII}), \tag{4.6.1}$$

while the absorption rate is

$$(RA)_0 = \frac{Q_I}{\Sigma_{aI}} \Sigma_{aII} V_{II} P_0(\Sigma_{TII}). \tag{4.6.2}$$

Here $G(\Sigma_{TII})$ and $P_0(\Sigma_{TII})$ are, respectively, region II sticking probabilities and escape probabilities *with all region II scattering replaced by absorption*. The subscript zero denotes an unperturbed reaction rate.

---

† From Eq. (4.4.11) we deduce that $G(\Sigma_{TII}) = \Sigma_{TII} \bar{l}_{II} P_0(\Sigma_{TII})$.

In Section 4.5 a single Monte Carlo was used to correct the unperturbed absorption rate. Here two Monte Carlo calculations will be required. First, we shall have to delete from the unperturbed entering flux the contributions of all neutrons which have collided in the lump. Then it will be necessary to reinstate contributions due to neutrons whose first collisions in the lump were scattering collisions, and not absorptions.

The deletion of collided neutrons proceeds as in Section 4.5. We simulate an outgoing surface source distribution, $Q_p^+(\omega)$, such that

$$Q_p^+(\omega) = \frac{\mu}{4\pi} \frac{Q_I}{\Sigma_{aI}} (1 - e^{-\Sigma_{TII}l}), \qquad \mu \equiv \omega \cdot \mathbf{n} > 0. \qquad (4.6.3)$$

When a surface source neutron re-enters the lump, it is allowed to scatter normally. If it is reabsorbed in the lump, the weight of the absorbed neutron is subtracted from the absorption count. In any case, its weighted track length in the lump is deposited, with a negative sign, in a track length bin. Let the deleted absorption rate, computed via this first Monte Carlo, be $-(RA)_1$.

Having deleted all collided neutrons, we must now replace those whose first collisions in the lump were scattering collisions. To this end we generate the unperturbed *ingoing* surface source, $Q_p^-(\omega)$, and the scattering collisions produced by this source. Since the unperturbed ingoing flux is equal to $Q_I/(4\pi\Sigma_{aI})$,

$$Q_p^-(\omega) \equiv \frac{|\omega \cdot \mathbf{n}|}{4\pi} \frac{Q_I}{\Sigma_{aI}} = \frac{\mu Q_1}{4\pi\Sigma_{aI}}, \qquad \omega \cdot \mathbf{n} \equiv -\mu < 0. \qquad (4.6.4)$$

It is easy enough to simulate such a source. We suppose, by way of illustration, that starting points are chosen from a distribution uniform on the surface of the lump, and incoming ω's from an isotropic distribution. Arguing as in Section 4.3, we conclude that each starter must carry a weight equal to $\mu A_{II} Q_I / 2\Sigma_{aI}$.

We have, then, specified a starting procedure which generates the ingoing source. What do we know of the subsequent history of starters? We know that each starter collides in the lump before emerging from it. Its first collision in the lump *must* be a scattering collision, though all other collisions take place normally. By some device we must force such behavior on all starters, and we describe now how this may be done.

In Fig. 4.3 we depict a starter entering the lump at P_1 in the direction of ω. The fraction $\exp(-\Sigma_{TII}l)$ of the starter's weight escapes uncollided and is discarded. Correspondingly, we multiply the starter's weight by $[1 - \exp(-\Sigma_{TII}l)]$. A fragment of the starter remains and this fragment must collide somewhere on the chord P_1P_2. Now, the probability of a collision in the interval dx is, clearly, proportional to $\exp(-\Sigma_{TII}x)\,dx$, and

$$\max(\Sigma_{TII}x) = \Sigma_{TII}l. \qquad (4.6.5)$$

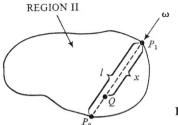

REGION II

Fig. 4.3

The quantity $\Sigma_{TII}x$ may, then, be drawn from a truncated exponential distribution by any of the methods described in Chapter 1 (see pp. 21, 35, for example). As usual, one should here avoid the use of conventional exponential routines, computing exponentials via g function tables (p. 151).

Having forced a first collision in the lump, we discard the absorbed fraction, $\Sigma_{aII}/\Sigma_{TII}$, of the particle's current weight and force the remaining fragment to scatter. At this point, we note, the particle has its final weight

$$W = \left(\frac{\mu A_{II}}{2}\frac{Q_{I}}{\Sigma_{aI}}\right)(1 - e^{-\Sigma_{TII}l})\left(\frac{\Sigma_{SII}}{\Sigma_{TII}}\right). \qquad (4.6.6)$$

If possible, we remove fluctuation in this weight by rejection, as in Section 4.3. It is easy to verify that

$$\overline{W} = \frac{1}{4}\frac{Q_{I}}{\Sigma_{aI}}\frac{\Sigma_{SII}}{\Sigma_{TII}}\Sigma_{TII}\bar{l}_{II}P_{0}(\Sigma_{TII})$$

$$= \frac{1}{4}\frac{Q_{I}}{\Sigma_{aI}}\frac{\Sigma_{SII}}{\Sigma_{TII}}G(\Sigma_{TII}), \qquad (4.6.7)$$

i.e. the average weight is equal to the rate of first scatterings in the lump. The histories of particles beyond their first collisions are treated by conventional Monte Carlo methods. Absorption rates may be estimated in various regions, as usual,[†] and we designate by $(RA)_2$ the contribution of our second Monte Carlo to the absorption rate in the lump. Since, in this second Monte Carlo, we are replacing previously deleted histories, *its* contribution will be positive. Summing all contributions we find[‡] that

$$RA = (RA)_0 - (RA)_1 + (RA)_2, \qquad (4.6.8)$$

where RA is the total absorption rate in region II.

---

† It should be noted that the track length laid down before the first collision should not be entered in the track length bin.

‡ Note the relation between Eqs. (4.6.8) and (4.3.7). If $\Sigma_{SII} = 0$, we find that $(RA)_0 = W$, $(RA)_1 = I$, $(RA)_2 = 0$, so that (4.6.8) and (4.3.7) become identical.

4.7 FIRST METHOD RE-EXAMINED

We have, in Section 4.6, been led by straightforward physical arguments to what we have called our first method for constructing a surface source in the presence of scattering. However, in order to prepare for future work, we think it worth while to develop this same method more formally. As before, we introduce an auxiliary problem, problem β. In problem β

$$Q_\beta(\mathbf{r}, \boldsymbol{\omega}) = Q_{II}/4\pi = \Sigma_{TII}Q_I/4\pi\Sigma_{aI}, \qquad \mathbf{r} \in R_{II},$$

$$Q_\beta(\mathbf{r}, \boldsymbol{\omega}) = 0, \qquad \mathbf{r} \in R_I.$$

Let

$$F_T(\mathbf{r}, \boldsymbol{\omega}) \equiv F_\alpha(\mathbf{r}, \boldsymbol{\omega}) + F_\beta(\mathbf{r}, \boldsymbol{\omega}),$$

$$\mathscr{L}_I F(\mathbf{r}, \boldsymbol{\omega}) \equiv \boldsymbol{\omega} \cdot \nabla F(\mathbf{r}, \boldsymbol{\omega}) + \Sigma_{TI}F(\mathbf{r}, \boldsymbol{\omega}),$$

$$\mathscr{L}_{II} F(\mathbf{r}, \boldsymbol{\omega}) \equiv \boldsymbol{\omega} \cdot \nabla F(\mathbf{r}, \boldsymbol{\omega}) + \Sigma_{TII}F(\mathbf{r}, \boldsymbol{\omega}).$$

Then

$$\mathscr{L}_I F_T(\mathbf{r}, \boldsymbol{\omega}) = \int d\boldsymbol{\omega}' \Sigma_{sI}(\boldsymbol{\omega} \cdot \boldsymbol{\omega}')F_T(\mathbf{r}, \boldsymbol{\omega}') + \frac{1}{4\pi}Q_I, \qquad \mathbf{r} \in R_I, \qquad (4.7.1)$$

$$\mathscr{L}_{II} F_T(\mathbf{r}, \boldsymbol{\omega}) = \int d\boldsymbol{\omega}' \Sigma_{sII}(\boldsymbol{\omega} \cdot \boldsymbol{\omega}')F_T(\mathbf{r}, \boldsymbol{\omega}') + \frac{1}{4\pi}Q_{II}, \qquad \mathbf{r} \in R_{II}. \qquad (4.7.2)$$

Further, $F_T(\mathbf{r}, \boldsymbol{\omega})$ satisfies reflecting conditions at the cell boundary. In fact, it will be assumed in this section that all fluxes fulfill such boundary conditions, unless other boundary conditions are specified.

Define a flux, $F_{TU}(\mathbf{r}, \boldsymbol{\omega})$, such that

$$\mathscr{L}_I F_{TU}(\mathbf{r}, \boldsymbol{\omega}) = \int d\boldsymbol{\omega}' \Sigma_{sI}(\boldsymbol{\omega} \cdot \boldsymbol{\omega}')F_{TU}(\mathbf{r}, \boldsymbol{\omega}) + \frac{1}{4\pi}Q_I, \qquad \mathbf{r} \in R_I, \qquad (4.7.3)$$

$$\mathscr{L}_{II} F_{TU}(\mathbf{r}, \boldsymbol{\omega}) = \frac{1}{4\pi}Q_{II}, \qquad \mathbf{r} \in R_{II}. \qquad (4.7.4)$$

Note that in Eq. (4.7.4) we have treated all collisions in region II as absorptions. The effective absorption cross section in region II is equal to Σ_{TII}. Since, by definition, $Q_I/\Sigma_{aI} = Q_{II}/\Sigma_{TII}$,

$$F_{TU}(\mathbf{r}, \boldsymbol{\omega}) = Q_I/4\pi\Sigma_{aI} = Q_{II}/4\pi\Sigma_{TII}. \qquad (4.7.5)$$

Now define $F_{TS}(\mathbf{r}, \boldsymbol{\omega})$ so that

$$\mathscr{L}_I F_{TS}(\mathbf{r}, \boldsymbol{\omega}) = \int d\boldsymbol{\omega}' \Sigma_{sI}(\boldsymbol{\omega} \cdot \boldsymbol{\omega}')F_{TS}(\mathbf{r}, \boldsymbol{\omega}'), \qquad \mathbf{r} \in R_I, \qquad (4.7.6)$$

$$\mathscr{L}_{II} F_{TS}(\mathbf{r}, \boldsymbol{\omega}) = \int d\boldsymbol{\omega}' \Sigma_{sII}(\boldsymbol{\omega} \cdot \boldsymbol{\omega}')F_{TS}(\mathbf{r}, \boldsymbol{\omega}')$$
$$+ \int d\boldsymbol{\omega}' \Sigma_{sII}(\boldsymbol{\omega} \cdot \boldsymbol{\omega}')F_{TU}(\mathbf{r}, \boldsymbol{\omega}'), \qquad \mathbf{r} \in R_{II}. \qquad (4.7.7)$$

Adding Eqs. (4.7.3) and (4.7.6) on the one hand, Eqs. (4.7.4) and (4.7.7) on the other, we see that

$$F_T(\mathbf{r}, \omega) = F_{TU}(\mathbf{r}, \omega) + F_{TS}(\mathbf{r}, \omega). \tag{4.7.8}$$

It should be clear that the above equation has a simple physical interpretation. We have, in Eq. (4.7.8), written $F_T(\mathbf{r}, \omega)$ as a sum of two fluxes. The flux $F_{TU}(\mathbf{r}, \omega)$ is generated by neutrons which have never scattered in the fuel. Neutrons which have scattered at least once in the fuel produce the second flux component, namely, the flux $F_{TS}(\mathbf{r}, \omega)$. From Eq. (4.7.8) we see that

$$F_\alpha(\mathbf{r}, \omega) + F_\beta(\mathbf{r}, \omega) = \frac{1}{4\pi} \frac{Q_{II}}{\Sigma_{TII}} + F_{TS}(\mathbf{r}, \omega),$$

$$F_\alpha(\mathbf{r}, \omega) = \frac{1}{4\pi} \frac{Q_{II}}{\Sigma_{TII}} + F_{TS}(\mathbf{r}, \omega) - F_\beta(\mathbf{r}, \omega). \tag{4.7.9}$$

At this point we separate F_β into components generated by different classes of neutrons. To accomplish the desired separation it will, unfortunately, be necessary to introduce some awkward terminology. We say that a neutron in the fuel is an "unreturned" neutron if it was born in the fuel and never left it. It is a "returned" neutron if it was born in the fuel, moved out of the fuel at least once, and subsequently re-entered the fuel. A neutron carries the label "unscattered before exit" if it was born in the fuel and was absorbed in or emitted from the fuel without first being scattered. If it was born in the fuel, and scattered *before* absorption or emission, it is labelled "scattered before exit". Thus, in the fuel†

$$F_\beta = F_\beta \text{ (unscattered before exit, unreturned)}$$

$$+ F_\beta \text{ (unscattered before exit, returned)}$$

$$+ F_\beta \text{ (scattered before exit)}.$$

It is clear that we shall have to abbreviate our notation somewhat, so we shall write

$$F_\beta \text{ (unscattered before exit, unreturned)} \equiv F_{\beta uu},$$

$$F_\beta \text{ (unscattered before exit, returned)} \equiv F_{\beta ur},$$

$$F_\beta \text{ (scattered before exit)} \equiv F_{\beta s},$$

$$F_\beta = F_{\beta uu} + F_{\beta ur} + F_{\beta s}.$$

In our new notation

$$F_\alpha = \frac{1}{4\pi} \frac{Q_{II}}{\Sigma_{TII}} + F_{TS} - F_{\beta uu} - F_{\beta ur} - F_{\beta s}, \qquad \mathbf{r} \in R_{II}. \tag{4.7.10}$$

---

† The arguments $\mathbf{r}$ and ω have been suppressed here.

As in Section 4.6, we now set out to evaluate the absorption rate

$$RA \equiv \int_{II} dV \int d\omega' \Sigma_{aII} F_\alpha(\mathbf{r}, \omega') \equiv \int_{II} \Sigma_{aII}\phi_\alpha(\mathbf{r}) \, dV. \qquad (4.7.11)$$

It will be convenient, before proceeding further, to split off that part of RA which we shall compute deterministically. Let

$$I_0 = \int_{II} d\mathbf{r} \int d\omega' \Sigma_{aII} \left[\frac{1}{4\pi} \frac{Q_{II}}{\Sigma_{TII}} - F_{\beta uu} \right]$$

$$= \frac{\Sigma_{aII}}{\Sigma_{TII}} \int_{II} [Q_{II} - \Sigma_{TII}\phi_{\beta uu}] \, d\mathbf{r}. \qquad (4.7.12)$$

From the definition of $F_{\beta uu}$ we see that

$$\mathscr{L}_{II} F_{\beta uu}(\mathbf{r}, \omega) = \frac{1}{4\pi} Q_{II}, \qquad \mathbf{r} \in R_{II}, \qquad (4.7.13)$$

$$F_{\beta uu}(\mathbf{r}_p, \omega) = 0, \qquad \omega \cdot \mathbf{n} < 0. \qquad (4.7.14)$$

Here, as usual, $\mathbf{r}_p$ is the radius vector to an arbitrary point on τ, the surface of R_{II}. It is evident from Eqs. (4.7.12), (4.7.13), and (4.7.14) that

$$\int_{II} [Q_{II} - \Sigma_{TII}\phi_{\beta uu}] \, d\mathbf{r}$$

is the leakage rate of uncollided neutrons from R_{II}. Therefore,

$$I_0 = Q_{II} V_{II} P_{0(II)}(\Sigma_{TII}) \Sigma_{aII}/\Sigma_{TII}$$

$$= Q_I V_{II} P_{0(II)}(\Sigma_{TII}) \Sigma_{aII}/\Sigma_{aI}. \qquad (4.7.15)$$

Comparing Eqs. (4.7.15) and (4.6.2), we find that $I_0 = (RA)_0$, so that

$$RA = (RA)_0 - \int_{II} \Sigma_{aII}\phi_{\beta ur} \, d\mathbf{r} + \int_{II} \Sigma_{aII}(\phi_{TS} - \phi_{\beta s}) \, d\mathbf{r}. \qquad (4.7.16)$$

Next we define

$$I_1 = \int_{II} \Sigma_{aII}\phi_{\beta ur} \, d\mathbf{r}.$$

Recall that $\phi_{\beta ur}$ is the scalar flux produced in the fuel by neutrons which emerge from the fuel uncollided. But neutrons which leave the fuel uncollided may be drawn from the surface source†

$$Q_p^+(\omega) = \frac{\omega \cdot \mathbf{n}}{4\pi} \frac{Q_I}{\Sigma_{aI}} (1 - e^{-\Sigma_{TII}l}), \qquad \omega \cdot \mathbf{n} \geq 0,$$

$$Q_p^+(\omega) = 0, \qquad \omega \cdot \mathbf{n} < 0.$$

---

† See Section 4.3.

This is the same source we have used, earlier, in the computation of $(RA)_1$. It follows that $I_1 = (RA)_1$, and

$$RA = (RA)_0 - (RA)_1 + \int_{II} \Sigma_{aII}(\phi_{TS} - \phi_{\beta s})\, d\mathbf{r}. \qquad (4.7.17)$$

Finally, we prove that

$$I_2 \equiv \int_{II} \Sigma_{aII}(\phi_{TS} - \phi_{\beta s})\, d\mathbf{r} = (RA)_2.$$

If we define $F_{TS} - F_{\beta s} \equiv F_{\Delta s}$, $\phi_{TS} - \phi_{\beta s} = \phi_{\Delta s}$, then

$$I_2 = \int_{II} \Sigma_{aII} \phi_{\Delta s}\, d\mathbf{r}. \qquad (4.7.18)$$

To show that I_2 and $(RA)_2$ are equal, it will be necessary to examine, rather closely, the properties of F_{TS} and $F_{\beta s}$. By definition,

$$\mathscr{L}_I F_{\beta s}(\mathbf{r}, \boldsymbol{\omega}) = \int d\boldsymbol{\omega}' \Sigma_{sI}(\boldsymbol{\omega} \cdot \boldsymbol{\omega}') F_{\beta s}(\mathbf{r}, \boldsymbol{\omega}'), \qquad \mathbf{r} \in R_I, \qquad (4.7.19)$$

$$\mathscr{L}_{II} F_{\beta s}(\mathbf{r}, \boldsymbol{\omega}) = \int d\boldsymbol{\omega}' \Sigma_{sII}(\boldsymbol{\omega} \cdot \boldsymbol{\omega}') F_{\beta s}(\mathbf{r}, \boldsymbol{\omega}')$$

$$+ \int d\boldsymbol{\omega}' \Sigma_{sII}(\boldsymbol{\omega} \cdot \boldsymbol{\omega}') F_{\beta uu}(\mathbf{r}, \boldsymbol{\omega}'), \qquad \mathbf{r} \in R_{II}. \quad (4.7.20)$$

We already know (see Eqs. 4.7.6 and 4.7.7) that

$$\mathscr{L}_I F_{TS}(\mathbf{r}, \boldsymbol{\omega}) = \int d\boldsymbol{\omega}' \Sigma_{sI}(\boldsymbol{\omega} \cdot \boldsymbol{\omega}') F_{TS}(\mathbf{r}, \boldsymbol{\omega}'), \qquad \mathbf{r} \in R_I,$$

$$\mathscr{L}_{II} F_{TS}(\mathbf{r}, \boldsymbol{\omega}) = \int d\boldsymbol{\omega}' \Sigma_{sII}(\boldsymbol{\omega} \cdot \boldsymbol{\omega}') F_{TS}(\mathbf{r}, \boldsymbol{\omega}')$$

$$+ \int d\boldsymbol{\omega}' \Sigma_{sII}(\boldsymbol{\omega} \cdot \boldsymbol{\omega}') F_{TU}(\mathbf{r}, \boldsymbol{\omega}), \qquad \mathbf{r} \in R_{II}.$$

Thus,

$$\mathscr{L}_I F_{\Delta s}(\mathbf{r}, \boldsymbol{\omega}) = \int d\boldsymbol{\omega}' \Sigma_{sI}(\boldsymbol{\omega} \cdot \boldsymbol{\omega}') F_{\Delta s}(\mathbf{r}, \boldsymbol{\omega}'), \qquad \mathbf{r} \in R_I, \qquad (4.7.21)$$

$$\mathscr{L}_{II} F_{\Delta s}(\mathbf{r}, \boldsymbol{\omega}) = \int d\boldsymbol{\omega}' \Sigma_{sII}(\boldsymbol{\omega} \cdot \boldsymbol{\omega}') F_{\Delta s}(\mathbf{r}, \boldsymbol{\omega}')$$

$$+ \int d\boldsymbol{\omega}' \Sigma_{sII}(\boldsymbol{\omega} \cdot \boldsymbol{\omega}')[F_{TU}(\mathbf{r}, \boldsymbol{\omega}') - F_{\beta uu}(\mathbf{r}, \boldsymbol{\omega}')], \qquad \mathbf{r} \in R_{II}. \quad (4.7.22)$$

Now, for $\mathbf{r} \in R_{II}$, the flux F_{TU} is determined by the equations

$$\mathscr{L}_{II} F_{TU}(\mathbf{r}, \boldsymbol{\omega}) = \frac{1}{4\pi} Q_{II}, \tag{4.7.23}$$

$$F_{TU}(\mathbf{r}_p, \boldsymbol{\omega}) = \frac{1}{4\pi} \frac{Q_{II}}{\Sigma_{TII}} = \frac{1}{4\pi} \frac{Q_{I}}{\Sigma_{aI}}, \qquad \boldsymbol{\omega} \cdot \mathbf{n} < 0. \tag{4.7.24}$$

Similarly,

$$\mathscr{L}_{II} F_{\beta uu}(\mathbf{r}, \boldsymbol{\omega}) = \frac{1}{4\pi} Q_{II}, \qquad \mathbf{r} \in R_{II}, \tag{4.7.25}$$

$$F_{\beta uu}(\mathbf{r}_p, \boldsymbol{\omega}) = 0, \qquad \boldsymbol{\omega} \cdot \mathbf{n} < 0. \tag{4.7.26}$$

Therefore,

$$\mathscr{L}_{II}[F_{TU} - F_{\beta uu}] = 0, \qquad \mathbf{r} \in R_{II}, \tag{4.7.27}$$

$$F_{TU}(\mathbf{r}_p, \boldsymbol{\omega}) - F_{\beta uu}(\mathbf{r}_p, \boldsymbol{\omega}) = \frac{1}{4\pi} \frac{Q_{II}}{\Sigma_{TII}} = \frac{1}{4\pi} \frac{Q_{I}}{\Sigma_{aI}}, \qquad \boldsymbol{\omega} \cdot \mathbf{n} < 0. \tag{4.7.28}$$

It is clear that the quantity $[F_{TU}(\mathbf{r}, \boldsymbol{\omega}) - F_{\beta uu}(\mathbf{r}, \boldsymbol{\omega})]$ satisfies a transport equation in the fuel, and that this quantity can be regarded as a flux produced by a surface source.

Perhaps it will be useful to pause here to examine the results of our analysis. From Eq. (4.7.18) we see that I_2 is the absorption rate induced in the fuel by the flux $F_{\Delta s}$. It follows from Eqs. (4.7.21) and (4.7.22) that $F_{\Delta s}$ is produced by neutrons scattered, within R_{II}, in the flux $F_{TU} - F_{\beta uu}$. To compute I_2 we must

a) generate the flux $F_{TU} - F_{\beta uu}$,
b) generate scattering events in this flux,
c) track the scattered neutrons, and, finally,
d) estimate the rate at which these scattered neutrons are absorbed in the fuel.

Given Eqs. (4.7.27) and (4.7.28) it is easy to carry out step (a). We draw starting points from a distribution uniform on τ and incoming $\boldsymbol{\omega}$'s isotropically. Each starter is given the weight

$$\tilde{W} = |\boldsymbol{\omega} \cdot \mathbf{n}| A_{II} Q_{I}(1 - e^{-\Sigma_{TII} l})/2\Sigma_{aI}.$$

Path lengths, x, to first collision points are drawn from the density

$$P(x) = \Sigma_{TII} e^{-\Sigma_{TII} x}/(1 - e^{-\Sigma_{TII} l}), \qquad 0 \leq x \leq l.$$

To carry out step (b) we multiply each weight by $\Sigma_{sII}/\Sigma_{TII}$. Each starter now carries the weight

$$W = |\boldsymbol{\omega} \cdot \mathbf{n}| A_{II} Q_{I} \Sigma_{sII}(1 - e^{-\Sigma_{TII} l})/2\Sigma_{aI}\Sigma_{TII}:$$

all first collisions are forced to be scattering collisions. From this point on, the remainder of the Monte Carlo calculation is perfectly conventional. It will be seen that the Monte Carlo process just outlined is the same as the process used in Section 4.6 to compute $(RA)_2$, so that indeed $I_2 = (RA)_2$ and, again,

$$RA = (RA)_0 - (RA)_1 + (RA)_2.$$

It is important to note that the procedure we have just outlined has a serious defect, one that is not easily remedied. By definition,

$$(RA)_0 = Q_I \Sigma_{aII} A_{II} G(\Sigma_{TII})/4\Sigma_{aI}\Sigma_{TII}.$$

One can show that $\overline{W}_1$, the average weight of a starter in the first Monte Carlo, is given by the expression

$$\overline{W}_1 = Q_I A_{II} G(\Sigma_{TII})/4\Sigma_{aI},$$

while

$$\overline{W}_2 = Q_I \Sigma_{sII} A_{II} G(\Sigma_{TII})/4\Sigma_{aI}\Sigma_{TII}.$$

When $\Sigma_{aII} \ll \Sigma_{TII}$, $\overline{W}_1 \doteq \overline{W}_2 \gg (RA)_0$. Thus, if there is a substantial probability that starters will eventually be absorbed in the fuel, we may find that

$$|RA - (RA)_0| \doteq (RA)_0,$$

or even that

$$RA - (RA)_0 \gg (RA)_0.$$

In either case $(RA)_0$ is only a small part of the absorption rate in the fuel, and we gain little by computing $(RA)_0$ deterministically. Still worse, it may happen that

$$(RA)_1 \doteq (RA)_2 \gg (RA)_0.$$

Should this be true, the results of the two Monte Carlo's will almost cancel. Consequently, the variance in the difference $(RA)_2 - (RA)_1$ may be large compared to the difference itself, and very large compared to the net absorption rate. We conclude that our first method is useful only when $\Sigma_{aII} \gg \Sigma_{sII}$.

4.8 SECOND METHOD

We describe now a second method which, in some respects, is simpler than the first. Again we introduce a problem β, but here†

$$Q_\beta(\mathbf{r}, \boldsymbol{\omega}) = \begin{cases} \Sigma_{aII}Q_I/4\pi\Sigma_{aI} \equiv Q_{II}/4\pi, & \mathbf{r} \in R_{II}, \\ 0, & \mathbf{r} \in R_I, \end{cases} \qquad (4.8.1)$$

---

† See, for comparison, the first equation in Section 4.7.

just as in Section 4.2.   Therefore, as in Section 4.2,

$$RA \equiv \int_{II} \Sigma_{aII} \phi_\alpha(\mathbf{r})\, dV = Q_{II} V_{II} - \int_{II} \Sigma_{aII} \phi_\beta(\mathbf{r})\, dV.$$

In the notation of Section 4.7

$$\phi_\beta = \phi_{\beta uu} + \phi_{\beta ur} + \phi_{\beta s},$$

and

$$RA = \left[Q_{II} V_{II} - \int_{II} \Sigma_{aII} \phi_{\beta uu}(\mathbf{r})\, d\mathbf{r} \right] - \int_{II} \Sigma_{aII} \phi_{\beta u}(\mathbf{r})\, d\mathbf{r} - \int_{II} \Sigma_{aII} \phi_{\beta s}(\mathbf{r})\, d\mathbf{r}.$$

Recall that

$$\mathscr{L}_{II} F_{\beta uu}(\mathbf{r}, \boldsymbol{\omega}) = \frac{1}{4\pi} Q_{II}, \qquad \mathbf{r} \in R_{II}, \tag{4.7.13}$$

while

$$F_{\beta uu}(\mathbf{r}_p, \boldsymbol{\omega}) = 0, \qquad \boldsymbol{\omega} \cdot \mathbf{n} < 0.$$

It follows that

$$\int_{II} \Sigma_{TII} \phi_{\beta uu}(\mathbf{r})\, d\mathbf{r} = Q_{II} V_{II} [1 - P_0(\Sigma_{TII})],$$

$$\begin{aligned}
(ra)_0 &\equiv Q_{II} V_{II} - \int_{II} \Sigma_{aII} \phi_{\beta uu}(\mathbf{r})\, d\mathbf{r} \\
&= Q_{II} V_{II} \left\{ P_0(\Sigma_{TII}) + \frac{\Sigma_{sII}}{\Sigma_{TII}} [1 - P_0(\Sigma_{TII})] \right\} \\
&= \frac{Q_I V_{II}}{\Sigma_{aI}} \Sigma_{aII} \left\{ P_0(\Sigma_{TII}) + \frac{\Sigma_{sII}}{\Sigma_{TII}} [1 - P_0(\Sigma_{TII})] \right\}. \tag{4.8.2}
\end{aligned}$$

We see that it is easy to evaluate $(ra)_0$ if $P_0(\Sigma_{TII})$ is known.   Defining

$$J_1 = \int_{II} \Sigma_{aII} \phi_{\beta ur}(\mathbf{r})\, d\mathbf{r}, \qquad J_2 = \int_{II} \Sigma_{aII} \phi_{\beta s}(\mathbf{r})\, d\mathbf{r},$$

we may write

$$RA = (ra)_0 - J_1 - J_2. \tag{4.8.3}$$

To compute J_1, we draw starters from the surface source

$$\begin{aligned}
q_p^+(\boldsymbol{\omega}) &= \boldsymbol{\omega} \cdot \mathbf{n} Q_{II}(1 - e^{-\Sigma_{TII} l})/4\pi \Sigma_{TII}, \qquad \boldsymbol{\omega} \cdot \mathbf{n} > 0, \\
&= \boldsymbol{\omega} \cdot \mathbf{n} Q_I \Sigma_{aII}(1 - e^{-\Sigma_{TII} l})/4\pi \Sigma_{aI} \Sigma_{TII}, \qquad \boldsymbol{\omega} \cdot \mathbf{n} > 0, \\
q_p^+(\boldsymbol{\omega}) &= 0, \qquad \boldsymbol{\omega} \cdot \mathbf{n} \le 0,
\end{aligned}$$

and generate their histories by conventional methods.   The process of sampling a surface source is by now familar to the reader and we say no more about it.

The computation of J_2 may be carried out in many ways: we choose a procedure which is completely straightforward. Since Q_{II} is uniform and isotropic, we choose starting points from a distribution uniform over V_{II} and draw ω's from an isotropic distribution. If a starter leaves V_{II} on its first flight, it is discarded. Otherwise its first collision is regarded as a scattering collision and it is assigned the weight

$$W_2 = Q_{II}V_{II}\Sigma_{sII}[1 - P_0(\Sigma_{TII})]/\Sigma_{TII}.$$

The remainder of the sample history evolves conventionally. By conventional means we compute a sample contribution to the absorption rate produced, in V_{II}, by each starter. It is clear that, if $P_0(\Sigma_{TII})$ is close to one, many starters will be rejected and our sampling process will be inefficient. However, as we shall see in Section 6.5, this inefficiency is not as serious a problem in multi-energy calculations as it is in one-energy calculations.

We shall not dwell at length on the performance of the second method, in one-group calculations. There are, however, two features of this second method which will prove important in the work of Chapter 6. First, it is significant that J_1 and J_2 occur in (4.8.3) with negative signs. Therefore, if $(ra)_0 \doteq RA$, it must be true that $J_1 \ll RA$ and $J_2 \ll RA$. In other words, if $(ra)_0$, which is computed analytically, is approximately equal to RA, then both Monte Carlo corrections must be relatively small. In contrast, one may find, when using the first method, that $(RA)_0 \doteq RA$, while $(RA)_1 \doteq (RA)_2 \gg RA$. Second, it is noteworthy that Eq. (4.8.2) is similar in form to the expression for the resonance absorption rate in the narrow resonance approximation (Ref. 2). In Section 6.5 we shall use the narrow resonance formula to estimate the absorption rate, and correct the narrow resonance approximation by Monte Carlo. To formulate the Monte Carlo corrections, we shall make use of the ideas and techniques incorporated in our second method.

4.9 OTHER APPLICATIONS OF SUPERPOSITION

In preceding sections we have used superposition to solve a very simple one-energy problem. Our discussion of this problem had one main purpose. We wished to introduce ideas which would be helpful later, in Chapters 5 and 6. However, in serving this purpose we have, perhaps, given the reader a narrow view of the utility of the superposition principle. This principle may be helpful in various situations. To demonstrate its versatility we turn now to the problem configuration depicted in Fig. 4.4. Region I in Fig. 4.4 is a water channel, while regions II and III are fuel regions, identical in composition. We will assume that the source in region I (a slowing down source) is large enough so that

$$\frac{Q_I}{\Sigma_{aI}} > \frac{Q_{II}}{\Sigma_{aII}} = \frac{Q_{III}}{\Sigma_{aIII}} = \frac{Q_f}{\Sigma_{af}}. \qquad (4.9.1)$$

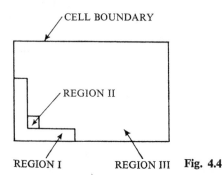

REGION I   REGION III   **Fig. 4.4**

Here Q_f and Σ_{af} are, respectively, the source and absorption cross section in the fuel. It is required that we compute the average flux, $\bar{\phi}_{II}$, in region II.

Of course this can be done by conventional Monte Carlo, but if region II is sufficiently small, such a Monte Carlo calculation will be very time consuming. Only a small fraction of the source neutrons will ever enter region II and many histories will be required to achieve a reasonably small variance.

Now let us imagine that the flux in the cell consists of two components,

$$\phi = \phi_a + \phi_b, \tag{4.9.2}$$

produced by corresponding components of the source, which we also split in two:

$$Q = Q_a + Q_b. \tag{4.9.3}$$

Specifically, let

$$Q_{aII} = Q_{aIII} = Q_f \tag{4.9.4}$$

and

$$Q_{aI} = Q_f \frac{\Sigma_{aI}}{\Sigma_{af}}. \tag{4.9.5}$$

As for the other component of the source, we have

$$Q_{bII} = Q_{bIII} = 0, \tag{4.9.6}$$

$$Q_{bI} = Q_I - Q_f \frac{\Sigma_{aI}}{\Sigma_{af}}. \tag{4.9.7}$$

Equation (4.9.1) shows that Q_{bI} is positive, so that both ϕ_a and ϕ_b are positive. In fact ϕ_a is flat and isotropic, and one can prove that

$$\phi_a = \frac{Q_{aI}}{\Sigma_{aI}} = \frac{Q_{aII}}{\Sigma_{aII}} = \frac{Q_{aIII}}{\Sigma_{aIII}}. \tag{4.9.8}$$

Thus, ϕ_a is known, and to determine ϕ we need only compute ϕ_b. We do this by conventional Monte Carlo.

The procedure outlined above has two main advantages over a straight-forward calculation of $\bar{\phi}_{II}$. First, Monte Carlo is used to compute only part of $\bar{\phi}_{II}$. Here again we see that superposition leads to a perturbation calculation. The flux ϕ_a is simply the flux level far from the water and the entire perturbing effect of the water is contained in ϕ_b. In one practical case, which we cite as an example, we have found that $\bar{\phi}_{bII} \doteq \frac{1}{2}\bar{\phi}_{II}$. Now, the variance in $\bar{\phi}_{aII}$ is zero. Therefore, in this case the variance, σ^2, in $\bar{\phi}_{II}$ is the same as the variance in $\bar{\phi}_{bII}$. It follows that

$$\frac{\sigma}{\bar{\phi}_{II}} \doteq \frac{1}{2} \frac{\sigma}{\bar{\phi}_{bII}}, \tag{4.9.9}$$

i.e. the relative error in $\bar{\phi}_{II}$ is only half as great as the relative error in $\bar{\phi}_{bII}$. Because of this fact alone, the use of superposition, in such a situation, reduces the number of histories required to achieve a given accuracy by three-quarters.

In addition, however, we must note that the flux, ϕ_b, is produced by a source lying wholly in region I. No neutrons are started in the fuel, far from the region of interest. This distribution of source neutrons tends to increase efficiency in the Monte Carlo calculation of $\bar{\phi}_{bII}$. All in all, then, we may expect that the use of superposition in this case will reduce Monte Carlo running time considerably.

The problem which we set ourselves in Section 4.1 was rather different from the problem dealt with here. We have tried, in discussing these two problems, to suggest that superposition may be used in different ways. In fact, the superposition principle leads us, not to any single method, but to a set of tricks. We shall have more to say about the capabilities and limitations of superposition in Section 4.10, below. First, however, we shall consider another class of computations based, not on superposition, but on the reciprocity relation.

4.10 USE OF THE ADJOINT TRANSPORT EQUATION

We have in the preceding sections of this chapter discussed the use of super-position as a variance reducing device. With the aid of the superposition principle we have developed efficient methods for the computation of fluxes in small regions. However, the reader may recall that other techniques are available for the treatment of small regions, and that techniques based on reciprocity† have been devised specifically for this purpose. It is our intention here to examine rather closely the mechanics of the adjoint flux calculation, to explore applications of the reciprocity relations, and to compare salient features of adjoint and superposition methods.

---

† See Section 3.6 and Section 3.7.

For our purposes it will be convenient to work with the integro-differential form of the transport equation. We find it helpful, also, to derive the adjoint equation in this same form. Let a diffusing medium be contained in a region, V, enclosed by a reflecting boundary, C.† In this region, of course, the neutron flux is governed by the equation

$$\boldsymbol{\omega} \cdot \nabla F(\mathbf{r}, \boldsymbol{\omega}) + \Sigma_T(\mathbf{r}) F(\mathbf{r}, \boldsymbol{\omega}) - \int d\boldsymbol{\omega} \Sigma_s(\mathbf{r}, \boldsymbol{\omega} \cdot \boldsymbol{\omega}') F(\mathbf{r}, \boldsymbol{\omega}') = Q(\mathbf{r}, \boldsymbol{\omega}). \quad (4.10.1)$$

For brevity we write

$$TF(\mathbf{r}, \boldsymbol{\omega}) = Q(\mathbf{r}, \boldsymbol{\omega}), \qquad (4.10.2)$$

where T is the linear operator which appears on the left-hand side of (4.10.1). Note that

$$\phi(\mathbf{r} \cdot \boldsymbol{\omega}) \boldsymbol{\omega} \cdot \nabla \psi(\mathbf{r}, \boldsymbol{\omega}) + \psi(\mathbf{r}, \boldsymbol{\omega}) \boldsymbol{\omega} \cdot \nabla \phi(\mathbf{r}, \boldsymbol{\omega}) = \nabla \cdot \boldsymbol{\omega} \phi(\mathbf{r}, \boldsymbol{\omega}) \psi(\mathbf{r}, \boldsymbol{\omega}).$$

Therefore, by Green's theorem,

$$\int_V \phi(\mathbf{r}, \boldsymbol{\omega}) T \psi(\mathbf{r}, \boldsymbol{\omega}) \, d\boldsymbol{\omega} \, d\mathbf{r} - \int_V \psi(\mathbf{r}, \boldsymbol{\omega}) T^* \phi(\mathbf{r}, \boldsymbol{\omega}) \, d\boldsymbol{\omega} \, d\mathbf{r}$$
$$= \int_C \boldsymbol{\omega} \cdot \mathbf{n} \psi(\mathbf{r}, \boldsymbol{\omega}) \phi(\mathbf{r}, \boldsymbol{\omega}) \, d\boldsymbol{\omega} \, d\mathbf{r} \quad (4.10.3)$$

for all ψ and ϕ satisfying certain continuity conditions. Here T^* is an operator defined through the relation

$$T^* \equiv \left[-\boldsymbol{\omega} \cdot \nabla + \Sigma_T(\mathbf{r}) - \int d\boldsymbol{\omega}' \Sigma_s(\mathbf{r}, \boldsymbol{\omega} \cdot \boldsymbol{\omega}') \right], \qquad (4.10.4)$$

and often referred to as the operator adjoint to T.

Suppose that ψ and ϕ both satisfy reflecting boundary conditions on C. Then so does their product. Consequently,

$$\int_C \boldsymbol{\omega} \cdot \mathbf{n} \psi(\mathbf{r}, \boldsymbol{\omega}) \phi(\mathbf{r}, \boldsymbol{\omega}) \, d\boldsymbol{\omega} \, d\mathbf{r} = 0 \qquad (4.10.5)$$

and

$$\int_V \phi(\mathbf{r}, \boldsymbol{\omega}) T \psi(\mathbf{r}, \boldsymbol{\omega}) \, d\boldsymbol{\omega} \, d\mathbf{r} = \int_V \psi(\mathbf{r}, \boldsymbol{\omega}) T^* \phi(\mathbf{r}, \boldsymbol{\omega}) \, d\boldsymbol{\omega} \, d\mathbf{r}. \qquad (4.10.6)$$

It is precisely because of the validity of Eq. (4.10.6) that we may refer to T^* as the adjoint transport operator.

Now we introduce the adjoint transport equation

$$T^* F^*(\mathbf{r}, \boldsymbol{\omega}) = Q^*(\mathbf{r}, \boldsymbol{\omega}), \qquad (4.10.7)$$

---

† It is not difficult to deal with other boundary conditions, but little can be gained here by a more general treatment of the boundary.

without, for the moment, specifying Q^*. Multiply Eq. (4.10.2) by F^*, Eq. (4.10.7) by F, and integrate over $\boldsymbol{\omega}$ and V. Making use of (4.10.6), one finds that

$$\int_V Q^*(\mathbf{r}, \boldsymbol{\omega}) F(\mathbf{r}, \boldsymbol{\omega}) \, d\mathbf{r} \, d\boldsymbol{\omega} = \int_V Q(\mathbf{r}, \boldsymbol{\omega}) F^*(\mathbf{r}, \boldsymbol{\omega}) \, d\mathbf{r} \, d\boldsymbol{\omega}. \qquad (4.10.8)$$

The reader should note the relation between Eqs. (4.10.8) and (3.7.27) derived earlier from the integral form of the transport equation. For monoenergetic neutrons, Eq. (3.7.27) reads as follows:

$$\int_V g(\mathbf{r}, \boldsymbol{\omega}) \psi(\mathbf{r}, \boldsymbol{\omega}) \, d\mathbf{r} \, d\boldsymbol{\omega} = \int Q(\mathbf{r}, \boldsymbol{\omega}) \chi^*(\mathbf{r}, \boldsymbol{\omega}) \, d\mathbf{r} \, d\boldsymbol{\omega}. \qquad (3.7.27a)$$

If, in (3.7.27a), we let $g(\mathbf{r}, \boldsymbol{\omega}) = Q^*(\mathbf{r}, \boldsymbol{\omega})/\Sigma_T(\mathbf{r})$, then we find that

$$\int_V Q^*(\mathbf{r}, \boldsymbol{\omega}) F(\mathbf{r}, \boldsymbol{\omega}) \, d\mathbf{r} \, d\boldsymbol{\omega} = \int_V Q(\mathbf{r}, \boldsymbol{\omega}) \chi^*(\mathbf{r}, \boldsymbol{\omega}) \, d\mathbf{r} \, d\boldsymbol{\omega}. \qquad (3.7.27b)$$

From Eqs. (3.7.27b) and (4.10.8) it is clear that

$$\int_V Q(\mathbf{r}, \boldsymbol{\omega}) \chi^*(\mathbf{r}, \boldsymbol{\omega}) \, d\mathbf{r} \, d\boldsymbol{\omega} = \int_V Q(\mathbf{r}, \boldsymbol{\omega}) F^*(\mathbf{r}, \boldsymbol{\omega}) \, d\mathbf{r} \, d\boldsymbol{\omega} \qquad (3.7.27c)$$

for all $Q(\mathbf{r}, \boldsymbol{\omega})$, including $Q(\mathbf{r}, \boldsymbol{\omega}) = \dfrac{1}{4\pi} \delta(\mathbf{r}_0 - \mathbf{r})$. Thus, postulating that $g(\mathbf{r}, \boldsymbol{\omega}) \equiv Q^*(\mathbf{r}, \boldsymbol{\omega})/\Sigma_T(\mathbf{r})$, we conclude that $\chi^*(\mathbf{r}, \boldsymbol{\omega}) = F^*(\mathbf{r}, \boldsymbol{\omega})$, as was asserted in Chapter 3.

If, in region R (some region of the cell),

$$Q^*(\mathbf{r}, \boldsymbol{\omega}) = \frac{1}{4\pi} \Sigma_{aR} \qquad (4.10.9)$$

and if

$$Q^*(\mathbf{r}, \boldsymbol{\omega}) = 0 \qquad (4.10.10)$$

elsewhere, then

$$\frac{1}{4\pi} V_R \Sigma_{aR} \bar{\phi}_R = \int_V Q(\mathbf{r}, \boldsymbol{\omega}) F^*(\mathbf{r}, \boldsymbol{\omega}) \, d\mathbf{r} \, d\boldsymbol{\omega}. \qquad (4.10.11)$$

On the other hand, if

$$Q^*(\mathbf{r}, \boldsymbol{\omega}) = \frac{1}{4\pi} \delta(\mathbf{r} - \mathbf{r}_0), \qquad (4.10.12)$$

then

$$\frac{1}{4\pi} \phi(\mathbf{r}_0) = \int_V Q(\mathbf{r}, \boldsymbol{\omega}) F^*(\mathbf{r}, \boldsymbol{\omega}) \, d\mathbf{r} \, d\boldsymbol{\omega}. \qquad (4.10.13)$$

In order to make use of Eqs. (4.10.11) and (4.10.13), it is necessary, of course, to find the adjoint flux. But the Monte Carlo methods heretofore described are designed to solve the transport equation, not its adjoint.

Fortunately, in one-energy problems the transport and adjoint equations are essentially equivalent. For if, in the adjoint equation

$$-\omega \cdot \nabla F^*(\mathbf{r}, \omega) + \Sigma_T(\mathbf{r})F^*(\mathbf{r}, \omega) - \int d\omega' \Sigma_s(\mathbf{r}, \omega \cdot \omega')F^*(\mathbf{r}, \omega')$$
$$= Q^*(\mathbf{r}, \omega) \quad (4.10.14)$$

we make the substitution†

$$-\omega \equiv \omega_0, \qquad -\omega' \equiv \omega_0', \qquad F^*(\mathbf{r}, \omega) = F^*(\mathbf{r}, -\omega_0) = f^*(\mathbf{r}, \omega_0),$$
$$(4.10.15)$$

we find that

$$\omega_0 \cdot \nabla f^*(\mathbf{r}, \omega_0) + \Sigma_T(\mathbf{r})f^*(\mathbf{r}, \omega_0) - \int d\omega_0' \Sigma_s(\mathbf{r}, \omega_0 \cdot \omega_0')f^*(\mathbf{r}, \omega_0')$$
$$= Q^*(\mathbf{r}, -\omega_0). \quad (4.10.16)$$

Furthermore, if $F^*(\mathbf{r}, \omega)$ satisfies reflecting boundary conditions on C, then so also does $f^*(\mathbf{r}, \omega_0)$. Thus, any method designed to solve transport equations in cells will give us the function $f^*(\mathbf{r}, \omega_0)$ which is related to the adjoint flux through (4.10.15).

Often we are interested only in $\phi^*(\mathbf{r})$, the adjoint scalar flux:

$$\phi^*(\mathbf{r}) \equiv \int d\omega F^*(\mathbf{r}, \omega).$$

It is worth noting that

$$\phi^*(r) = \int d\omega f^*(\mathbf{r}, \omega),$$

i.e. the adjoint scalar flux is equal to the ordinary scalar flux produced in a cell by the source $Q^*(\mathbf{r}, -\omega)$. If

$$Q(\mathbf{r}, \omega) = \frac{1}{4\pi} \delta(\mathbf{r} - \mathbf{r}_1) \quad (4.10.17)$$

and

$$Q^*(\mathbf{r}, \omega) = \frac{1}{4\pi} \delta(\mathbf{r} - \mathbf{r}_2), \quad (4.10.18)$$

we see from (4.10.8) that

$$\phi^*(\mathbf{r}_1) = \phi(\mathbf{r}_2). \quad (4.10.19)$$

Equation (4.10.19) is an important reciprocity relation. It states that the scalar flux produced at point 2 by a unit isotropic source at point 1 is the same as the scalar flux at 1 due to a unit isotropic source at 2.

Reciprocity and superposition lead to two different sets of variance reduction techniques. Methods based on superposition and those which rely on reciprocity are similar in some respects, but not identical. To bring out

---

† Apparently this substitution was first suggested by D. S. Selengut (Ref. 5).

differences between them, we retrace our steps and turn again to problems dealt with earlier in this chapter. Consider first the two-region problem configuration depicted in Fig. 4.1. We solve this problem now with the aid of Eq. (4.10.11). Since, in region I, $Q(\mathbf{r}, \boldsymbol{\omega})$ is constant and isotropic, and since the source vanishes in region II,

$$V_{\mathrm{II}}\Sigma_{a\mathrm{II}}\bar{\phi}_{\mathrm{II}} = Q_{\mathrm{I}}V_{\mathrm{I}}\bar{\phi}_{\mathrm{I}}^* = V_{\mathrm{I}}\Sigma_{a\mathrm{I}}\left(\frac{Q_{\mathrm{I}}}{\Sigma_{a\mathrm{I}}}\,\bar{\phi}_{\mathrm{I}}^*\right). \tag{4.10.20}$$

Now, the adjoint flux in Eq. (4.10.20) is produced by a source of density equal to $\Sigma_{a\mathrm{II}}$. On the other hand, the flux, ϕ_β, in Eq. (4.2.5) is generated by the source Q_{II}:

$$Q_{\mathrm{II}} = \Sigma_{a\mathrm{II}}\frac{Q_{\mathrm{I}}}{\Sigma_{a\mathrm{I}}}. \tag{4.10.21}$$

Therefore,

$$\frac{Q_{\mathrm{I}}}{\Sigma_{a\mathrm{I}}}\bar{\phi}_{\mathrm{I}}^* = \bar{\phi}_{\beta\mathrm{I}}, \tag{4.10.22}$$

and

$$V_{\mathrm{II}}\Sigma_{a\mathrm{II}}\bar{\phi}_{\mathrm{II}} = V_{\mathrm{I}}\Sigma_{a\mathrm{I}}\bar{\phi}_{\beta\mathrm{I}}. \tag{4.10.23}$$

The reader will observe that Eqs. (4.2.5) and (4.10.23) are identical. But we have not deduced, from the reciprocity relation, an equivalent of Eq. (4.2.4). The superposition principle, in other words, leads to a relation between the flux *shapes* in problems α and β, while the reciprocity relation gives us much less information. Having solved the adjoint problem, we can, through Eq. (4.10.23), compute the *average flux* in region II, and no more. We see here signs of a fundamental weakness in methods based on reciprocity. Important advantages of such methods will soon be pointed out.

It should be clear that, in the one-energy setting of the present discussion, problem β and the adjoint problem are, in fact, the same problem. All the methods developed to solve problem β may be regarded, then, as tricks to use in adjoint calculations.

To bring out more sharply the differences between methods based on superposition and reciprocity, we re-examine our second problem (see Section 4.9). We have already solved this problem via the superposition principle. Clearly, we may, if we like, use reciprocity instead. Since we are interested in the region II absorption rate, we let

$$Q^* = \frac{1}{4\pi}\Sigma_{a\mathrm{II}} \tag{4.10.24}$$

in region II and

$$Q^* = 0 \tag{4.10.25}$$

elsewhere. It follows from (4.10.11) that

$$V_{\mathrm{II}}\Sigma_{a\mathrm{II}}\bar{\phi}_{\mathrm{II}} = V_{\mathrm{I}}Q_{\mathrm{I}}\bar{\phi}_{\mathrm{I}}^* + V_{\mathrm{II}}Q_{\mathrm{II}}\bar{\phi}_{\mathrm{II}}^* + V_{\mathrm{III}}Q_{\mathrm{III}}\bar{\phi}_{\mathrm{III}}^*.$$

Here we see that reciprocity and superposition have led to very different methods of solution. In using superposition we put a source in region I. On the other hand, the adjoint source was put in region II. Through superposition we computed only part of the region I absorption rate. Reciprocity gave us the whole absorption rate. A more basic difference between the two methods was noted earlier. It has already been pointed out that we may rely on superposition to give us flux *distributions*. Reciprocity, on the other hand, yields a *single* flux average in a prescribed region. However, in general, the adjoint method is much more powerful than superposition. Suppose, by way of illustration, that Q_{II} were smaller than Q_{III}. One could still, by superposition, eliminate Q_{III}. One could still construct a problem β with $Q_{III} = 0$, but in this problem Q_I would be positive and Q_{II} negative. The occurrence of sources with different signs in the same problem would tend to raise the variance, impairing the effectiveness of superposition. No such difficulty would arise in the adjoint calculation: it would proceed as before. Thus, a slight complication of the problem configuration may render the superposition method unfeasible without affecting the utility of the reciprocity relation.

We have deferred until now any discussion of the history of reciprocity and superposition methods in Monte Carlo. In so far as we can tell, superposition was first used as a Monte Carlo technique by the authors, and the approach was developed in 1958. The method has been applied to the calculation of thermal flux averages and also to the calculation of resonance escape probabilities (Refs. 6 and 7). The latter application will be described fully in Chapter 6. The history of the adjoint method is more complicated. It is certainly true that one of Albert's estimators (see Section 3.7) may be regarded as a conventional estimator applied to an adjoint integral equation. However, Albert himself did not point this out. In his paper he draws our attention to a specific method of estimation and not to the adjoint equation itself. It was Maynard who, in 1959 (Ref. 8), suggested explicitly that the use of the reciprocity relation might be advantageous in certain types of Monte Carlo calculations. Maynard's work deals directly with the adjoint transport equations, and it seems fair to say that the power and generality of the adjoint method was not brought out by Albert as it was by Maynard.

REFERENCES

1. K. M. CASE, F. DE HOFFMANN, and G. PLACZEK, *Introduction to the Theory of Neutron Diffusion*, Vol. I, Los Alamos Scientific Laboratory, Los Alamos, New Mexico (1953).

2. L. DRESNER, *Resonance Absorption in Nuclear Reactors*, Pergamon Press, New York (1960).

3. A. AMOUYAL, P. BENOIST, and J. HOROWITZ, "Nouvelle Methode de Determination du Facteur d'Utilisation Thermique d'Une Cellule," *J. Nucl. Energy*, 6, 79 (1957).

4. A. M. WEINBERG and E. P. WIGNER, *The Physical Theory of Neutron Chain Reactors*, Univ. of Chicago Press (1958).

5. D. S. SELENGUT, "Variational Analysis of Multi-Dimensional Systems," *Trans. Am. Nucl. Soc.*, **2**, No. 1, 58 (1959).

6. E. M. GELBARD, H. B. ONDIS, and J. SPANIER, "MARC—A Multigroup Monte Carlo Program for the Calculation of Capture Probabilities," *Bettis Atomic Power Laboratory*, *WAPD-TM*-273 (May, 1962).

7. B. L. ANDERSON, E. M. GELBARD, and J. SPANIER, "RESQ-2—A Combined Analytic Monte Carlo Calculation of Resonance Absorption Based on Superposition," *Bettis Atomic Power Laboratory*, *WAPD-TM*-665 (March, 1967).

8. C. W. MAYNARD, "An Application of the Reciprocity Theorem to the Acceleration of Monte Carlo Calculations," *Trans. Am. Nucl. Soc.*, **3**, No. 2 (1960).

5

Computation of Thermal Neutron Fluxes

5.1 INTRODUCTION

In Chapter 4 we have attempted, through a study of somewhat academic problems, to show the reader how the principles of superposition and reciprocity may be used in Monte Carlo calculations. Now we must turn our attention to more practical problems, to real problems encountered daily by reactor physicists and engineers. Below we shall discuss, in some detail, the multigroup computation of thermal neutron fluxes, concentrating specifically on techniques incorporated in the MARC program (Ref. 1). We consider first the ideas underlying the MARC formulation of multigroup equations, then deal briefly with conventional methods of simulation and estimation. However, we are primarily interested here in problems which cannot be solved satisfactorily by conventional methods, and we devote most of this chapter to the development of the adjoint method. Finally, in order to demonstrate the power of the adjoint method, we show how it may be combined with other techniques in the analysis of foil perturbations.

The reader should be aware of the fact that we are taking an eminently practical approach to reactor physics and design work. Because of the enormous capabilities of modern computers, it is now quite feasible to use Monte Carlo in routine design calculations. On the Philco-212 a typical MARC problem runs in about three minutes, and before this book appears in print much faster machines will be available.

5.2 STATEMENT OF THE THERMAL PROBLEM

In the design of thermal nuclear reactors the computation of the thermal neutron flux plays a most important role. Sometimes it is possible to calculate this flux by deterministic methods (Ref. 2), but often it is not. When the geometric structure of the reactor is complicated, one is usually forced to rely on Monte Carlo.

Two features of the thermal energy range distinguish it from the rest of the energy spectrum. One is upscattering—if a slow neutron strikes a nucleus in thermal motion, the neutron's energy may increase. On the other hand, if the neutron's energy is very much higher than kT, almost every scattering collision will slow it down. Thus, while we may ignore upscattering of fast neutrons, we cannot ignore the upscattering of thermal neutrons. In addition, we must, in the thermal range, take account of chemical binding. So long as the initial and final energies of a scattered neutron are high compared with the binding energy of the struck nucleus, binding effects may be neglected. The energy and angular distribution of scattered neutrons are then determined by simple kinematic laws. If either the initial or final energies are comparable with the binding energy, or smaller than the binding energy, however, binding effects become significant. Correspondingly, the scattering laws become extremely complicated.

Because one encounters special problems in the thermal range, difficulties peculiar to this energy band, it is customary to introduce a sharp, though artificial, boundary, E_c, between thermal and epithermal energies. For energies below E_c one postulates that

$$\boldsymbol{\omega} \cdot \nabla F(\mathbf{r}, E, \boldsymbol{\omega}) + \Sigma_t(\mathbf{r}, E)F(\mathbf{r}, E, \boldsymbol{\omega})$$

$$= \int_0^{E_c} \Sigma_t(\mathbf{r}, E')C(E', \boldsymbol{\omega}', E, \boldsymbol{\omega}; \mathbf{r})F(\mathbf{r}, E', \boldsymbol{\omega}')\, dE'\, d\boldsymbol{\omega}' + Q(\mathbf{r}, E, \boldsymbol{\omega}). \quad (5.2.1)$$

Here we have used the notation of Chapter 2 (Eq. 2.4.17) but we have separated the scalar energy E from the unit direction vector $\boldsymbol{\omega}$. The scattering kernel† $C(E', \boldsymbol{\omega}', E, \boldsymbol{\omega}; \mathbf{r})$ and the slowing-down density $Q(\mathbf{r}, E, \boldsymbol{\omega})$ are assumed to be known.

While it is not absolutely necessary, it is quite convenient, in formulating thermal multigroup equations, to make several important approximations. We suppose

a) that the slowing-down density is isotropic,

b) that only one element or compound acts as a moderator, and

c) that the slowing-down density is a separable function of energy and position, i.e.

$$Q(\mathbf{r}, E, \boldsymbol{\omega}) = \frac{1}{4\pi} \chi(E)q(\mathbf{r}).$$

Of course, we are principally concerned here with Monte Carlo, not with physics. Therefore, it would be inappropriate to discuss at length the validity of such approximations. Some brief comment seems necessary, nonetheless. Approximations (a), (b), and (c) are very often used in thermal neutron

---

† The computation of the kernel C is usually based on models which have been derived from theories such as Nelkin's (Ref. 3). Such theories, which are based on simplified treatments of binding effects, are incorporated in various computer programs (see, e.g. Ref. 2).

calculations. With regard to the accuracy of the first approximation, little is known; but there is some evidence that anisotropic slowing-down is *usually* unimportant. As for approximation (b), it is often true that one element in a reactor is much more effective as a moderator than all others combined. In this section we shall deal specifically with water-moderated reactors. Generally, in water-moderated reactors, hydrogen is by far the most powerful moderator. Particularly at low energies (i.e. at energies below 0.625 eV) moderation by other elements plays a minor role. It follows that assumption (c) is not unreasonable, for, in monatomic hydrogen, the isotropic slowing-down density, Q_H, is truly separable. In fact, one can show (Ref. 4) that

$$Q_H(\mathbf{r}, E) \equiv \int Q_H(\mathbf{r}, E, \boldsymbol{\omega}) \, d\boldsymbol{\omega} = q(\mathbf{r}) \operatorname{erf} \sqrt{E/kT}, \tag{5.2.2}$$

where T is the temperature of the hydrogen. In a first approximation, binding effects do not change the form of Eq. (5.2.2). Instead they raise the temperature which enters the equation, so that Eq. (5.2.2) becomes

$$Q_H(\mathbf{r}, E) = q(\mathbf{r}) \operatorname{erf} \sqrt{E/k\tilde{T}}. \tag{5.2.3}$$

Whereas, at room temperature, $kT = 0.025$ eV, the effective temperature yields $k\tilde{T} = 0.117$ eV (Ref. 3).

Since we have ignored the moderating properties of elements other than hydrogen (elements which we shall call "heavy elements"), we can gain little by preserving other relatively unimportant features of heavy element scattering. On the other hand, we can gain a lot by *further* simplifying our model of heavy elements, and we shall do so. In the work of this chapter we take the scattering cross sections of all heavy elements to be isotropic and energy independent, and for each element we let

$$\sigma_{sh} = (1 - \bar{\mu})\sigma_{s\infty}. \tag{5.2.4}$$

Here σ_{sh} is the assumed scattering cross section of the element, $\sigma_{s\infty}$ is the true epithermal scattering cross section,† and $\bar{\mu} = 2/(3A)$. The reader will recognize that our notation is conventional: $\bar{\mu}$ is the average cosine of the scattering angle for a free nucleus of mass M_h, while A is the ratio of M_h to the neutron mass. He will note also that Eq. (5.2.4) embodies the transport approximation, frequently used in reactor computations (Ref. 2). Any program which incorporates substantial approximations such as those above, cannot be regarded as an absolute standard of accuracy. It is, instead, a design tool, intended for heavy use by the engineer and physicist.

All of the approximations discussed above facilitate the Monte Carlo computation but, having made these approximations, we still have not

---

† This epithermal cross section is usually almost independent of energy, so that there is little ambiguity in the definition of $\sigma_{s\infty}$.

arrived at a tractable Monte Carlo problem. To simulate Eq. (5.2.1) directly it would be necessary, after each hydrogen scattering collision, to construct distribution functions for emergent energies and scattering angles.   In principle such a procedure is possible, but in practice the complexity of the scattering kernel makes it impractical. For this reason it seems to be necessary to precompute and store some sort of scattering matrix. In order to define scattering matrices, we must introduce a discrete energy mesh. This may be done in many ways—we describe first the approach used in the formulation of the MARC program.

In deriving the MARC equations one computes the flux at energy mesh points $E_1 \geq E_2 \geq \cdots \geq E_N$ and evaluates the scattering integral by the trapezoidal rule. As a result, one is led to a set of coupled transport equations

$$\boldsymbol{\omega} \cdot \nabla F_i(\mathbf{r}, \boldsymbol{\omega}) + \Sigma_{ti}(\mathbf{r}) F_i(\mathbf{r}, \boldsymbol{\omega}) = N_H(\mathbf{r}) \sum_{j=1}^{N} \int c_{ij}(\boldsymbol{\omega}', \boldsymbol{\omega}) W_j F_j(\mathbf{r}, \boldsymbol{\omega}') \, d\boldsymbol{\omega}'$$

$$+ \frac{\Sigma_{sh}(\mathbf{r})}{4\pi} \phi_i(\mathbf{r}) + \frac{\chi_i}{4\pi} q(\mathbf{r}), \qquad (5.2.5)$$

where $\Sigma_{ti}(\mathbf{r}) = \Sigma_{ai}(\mathbf{r}) + \Sigma_{sh}(\mathbf{r}) + \Sigma_{sHi}(\mathbf{r})$ is the sum of the absorption cross section for all the elements, the heavy element scattering cross section (Eq. 5.2.4), and the hydrogen scattering cross section. Of course,

$$\phi_i(\mathbf{r}) = \int F_i(\mathbf{r}, \boldsymbol{\omega}) \, d\boldsymbol{\omega}$$

is the scalar flux and the subscript i is used to designate the value of a function at energy E_i. Further,

$$N_H(\mathbf{r}) c_{ij}(\boldsymbol{\omega}', \boldsymbol{\omega}) = \Sigma_{tH}(\mathbf{r}, E_j) C_H(E_j, \boldsymbol{\omega}', E_i, \boldsymbol{\omega}; \mathbf{r}),$$

$$W_i = \frac{(E_{i-1} - E_i) + (E_i - E_{i+1})}{2} \equiv \frac{\Delta_{i-1} + \Delta_i}{2}, \qquad i \neq 1,$$

$$W_1 = \frac{E_1 - E_2}{2} = \Delta_1.$$

We shall take E_1 to be the upper (high energy) cut point of the thermal group (E_c in Eq. 5.2.1) and E_N to be the lowest energy in the mesh, $E_N \neq 0$. Since the flux at $E = 0$ vanishes, we permit no subscript in Eq. (5.2.5) to correspond to $E = 0$. In Eq. (5.2.5), $N_H(\mathbf{r})$ is the number density of the moderator, which we suppose to be hydrogen. Most of what we say, however, will be true of other moderators as well.

Before going further, we introduce a change of variable to simplify our future work. Let

$$\psi_i(\mathbf{r}, \boldsymbol{\omega}) = F_i(\mathbf{r}, \boldsymbol{\omega}) W_i, \qquad g_i(\mathbf{r}) = \phi_i(\mathbf{r}) W_i, \qquad B_i = \chi_i W_i.$$

Now we may write

$$\boldsymbol{\omega} \cdot \boldsymbol{\nabla}\psi_i(\mathbf{r}, \boldsymbol{\omega}) + \Sigma_{ti}(\mathbf{r})\psi_i(\mathbf{r}, \boldsymbol{\omega}) = N_H(\mathbf{r})\sum_{j=1}^{N} \int W_i c_{ij}(\boldsymbol{\omega}', \boldsymbol{\omega})\psi_j(\mathbf{r}, \boldsymbol{\omega}')\, d\boldsymbol{\omega}'$$

$$+ \frac{\Sigma_{sh}(\mathbf{r})}{4\pi}\, g_i(\mathbf{r}) + \frac{B_i}{4\pi}\, q(\mathbf{r}). \qquad (5.2.6)$$

The reader should note that the quantity $\Sigma_{ti}(\mathbf{r})$ is the sum of the total cross section of hydrogen and the transport cross sections of all heavy elements. In a sense, it is a total cross section at E_i. However, one can show that the solution of Eq. (5.2.6) in an infinite medium without absorption will not be Maxwellian. We therefore prefer to modify Eq. (5.2.6) so that this solution *will* be Maxwellian. To this end we define

$$\tilde{\sigma}_{sHi} \equiv \sum_{j=1}^{N} W_j \int c_{ji}(\boldsymbol{\omega}', \boldsymbol{\omega})\, d\boldsymbol{\omega}',$$

$$\tilde{\Sigma}_{sHi}(\mathbf{r}) = N_H(\mathbf{r})\tilde{\sigma}_{sHi},$$

and

$$\tilde{\Sigma}_{ti}(\mathbf{r}) = \Sigma_{ai}(\mathbf{r}) + \Sigma_{sh}(\mathbf{r}) + \tilde{\Sigma}_{sHi}(\mathbf{r}).$$

Introducing $\tilde{\Sigma}_{ti}(\mathbf{r})$ in place of Σ_{ti} in Eq. (5.2.6), we get

$$\boldsymbol{\omega} \cdot \boldsymbol{\nabla}\psi_i(\mathbf{r}, \boldsymbol{\omega}) + \tilde{\Sigma}_{ti}(\mathbf{r})\psi_i(\mathbf{r}, \boldsymbol{\omega}) = N_H(\mathbf{r})\sum_{j} \int W_i c_{ij}(\boldsymbol{\omega}', \boldsymbol{\omega})\psi_j(\mathbf{r}, \boldsymbol{\omega}')\, d\boldsymbol{\omega}'$$

$$+ \frac{\Sigma_{sh}(\mathbf{r})}{4\pi}\, g_i(\mathbf{r}) + \frac{B_i}{4\pi}\, q(\mathbf{r}). \qquad (5.2.7)$$

In an absorption-free infinite medium with no source, Eq. (5.2.7) takes the form

$$\tilde{\sigma}_{sHi}\phi_i = \sum_{j=1}^{N} \int c_{ij}(\boldsymbol{\omega}', \boldsymbol{\omega})W_j\phi_j\, d\boldsymbol{\omega}'. \qquad (5.2.8)$$

If we assume that ϕ is Maxwellian, $\phi_j = \phi_{Mj}$, then, by the detailed balance principle (Ref. 5),

$$\sum_{j=1}^{N} \int c_{ij}(\boldsymbol{\omega}', \boldsymbol{\omega})W_j\phi_{Mj}\, d\boldsymbol{\omega}' = \sum_{j=1}^{N} \int c_{ji}(\boldsymbol{\omega}', \boldsymbol{\omega})W_j\phi_{Mi}\, d'\omega'$$

$$= \tilde{\sigma}_{sHi}\phi_{Mi}.$$

Thus, a Maxwellian flux does satisfy Eq. (5.2.8). It is easy to show that $\tilde{\sigma}_{sHi} \doteq \sigma_{sHi}$, for

$$N_H(\mathbf{r})\tilde{\sigma}_{sHi} = \sum_{j=1}^{N} W_j \int \Sigma_{tH}(\mathbf{r}, E_i)C_H(E_i, \boldsymbol{\omega}', E_j, \boldsymbol{\omega}:\mathbf{r})\, d\boldsymbol{\omega}'. \qquad (5.2.9)$$

ay State Books

7509221
Saint Paul Street
th Smithfield, RI 02896

112-3425926-0325065

HIP TO:	VIA:
eremy Wittkopp	Order Number:
154 19861603662	Order Date: 11/10/2022
14 S 2ND AVE	Shipping Method: Standard Std US D2D
IECHANICVILLE, NY 12118-2221	Dom
Inited States	

SKU	QTY	TITLE	TOTAL
BSM.4NDC	1	Monte Carlo Principles and Neutron Transport Problems (Dover Books on Mathematics) [Paperback] [2008] Spanier, Jerome; Gelbard, Ely M.	10.62

Order Total: 11.36

ecial Instructions:

Interchanging ω and ω' and making use of the fact that C_H is a function only of the scalar product $\omega' \cdot \omega$,[†]

$$N_H(\mathbf{r})\tilde{\sigma}_{sHi} = \sum_{j=1} W_j \int \Sigma_{tH}(\mathbf{r}, E_i) C_H(E_i, \omega, E_j, \omega'; \mathbf{r})\, d\omega'$$

$$\doteq \int_0^{E_c} \int \Sigma_{tH}(\mathbf{r}, E_i) C_H(E_i, \omega, E', \omega'; \mathbf{r})\, d\omega'\, dE'$$

$$\equiv N_H(\mathbf{r})\sigma_{sHi}. \tag{5.2.10}$$

In fact,

$$\sum_{j=1}^{N} W_j C_H(E_i, \omega, E_j, \omega'; \mathbf{r})$$

is simply the trapezoidal integral, over final energies, of the scattering kernel for hydrogen.

So far we have said nothing about the dependence of the c_{ij} on the scattering angle. Any realistic scattering kernel will involve a very complicated function of the angle between incoming and outgoing neutron velocities. A finely detailed table of the kernel as a function of both energy *and* scattering angle would be very large. In fact, there seems to be no economical way, at present, to incorporate the true angular dependence of the scattering kernel in a Monte Carlo program. One is forced to deal with the scattering angle in some rough approximation.

Commonly, in deterministic transport calculations, the differential scattering cross section is expanded in a truncated series of Legendre polynomials. Approximately,[‡]

$$c_{ij}(\omega', \omega) = c_{ij}(\omega' \cdot \omega) = \frac{1}{2\pi} k_{ij}(\mu_0)$$

$$= \sum_{l=0}^{L} \left(\frac{2l+1}{4\pi} \right) k_{ijl} P_l(\mu_0).$$

Here $\mu_0 = \omega' \cdot \omega$, while P_l is the Legendre polynomial of order l. Unfortunately, the use of such an expansion causes difficulties in Monte Carlo simulation which are not encountered in deterministic computations. Near the thermal cut point, E_c, the diagonal matrix elements $k_{ii}(\mu_0)$ are strongly peaked about $\mu_0 = 0$. In other words, when there is little energy exchange we must be dealing with glancing collisions and scattering tends to be forward. As a result, the truncated expansion will be negative for some values of μ_0. If,

---

[†] This is not always true but exceptions to this rule are rare.
[‡] Note that $c_{ij}(\omega' \cdot \omega)\, d\omega$ is the differential scattering cross section for scattering into the solid angle $d\omega$, while $k_{ij}(\mu_0)\, d\mu_0$ is the differential cross section for scattering into an interval $d\mu_0$. The interval $d\mu_0$ subtends a solid angle $2\pi\, d\mu_0$.

for example, we treat the delta function $\delta(\mu_0)$ in a P_1-approximation (i.e. if we set $L = 1$ in the truncated expansion), we find that

$$\delta(\mu_0) \doteq \tfrac{1}{2} + \tfrac{3}{2}\mu_0 = \delta_1(\mu_0).$$

Obviously, $\delta_1(\mu_0)$ is negative when $\mu_0 < -\tfrac{1}{3}$. To simulate scattering by a kernel which is negative for some scattering angles, we must endow some neutrons with negative weights. Neutron weights will fluctuate in sign during the course of the calculation, and such fluctuations may have a catastrophic effect on the variance of statistical estimates of flux integrals.† For this reason it seems undesirable to use truncated P_l expansions of the kernel in Monte Carlo,‡ and particularly undesirable to use low-order P_l expansions. What, then, can one do? There are many ways to avoid the introduction of negative cross sections. We consider first a simple scheme used in the original version of MARC.

Define

$$a_{ij} = W_i \int_{-1}^{1} k_{ij}(\mu_0)\, d\mu_0,$$

$$\bar{\mu}_{ij}a_{ij} = W_i \int_{-1}^{1} \mu_0 k_{ij}(\mu_0)\, d\mu_0.$$

Since $k_{ij}(\mu_0) \geq 0$, it is clear that $|\bar{\mu}_{ij}| \leq 1$. Now suppose that a sample neutron at energy E_J has just been scattered by hydrogen. Let

$$p_{iJ} = a_{iJ} \Big/ \sum_{k=1}^{N} a_{kJ}$$

$$= a_{iJ}/\tilde{\sigma}_{sHJ}$$

define the discrete probability density for final energies E_i. Choose a final energy, say E_I, from this distribution. Now we postulate that, with probability $1 - |\bar{\mu}_{IJ}|$, the scattering is isotropic and with probability $|\bar{\mu}_{IJ}|$ the scattering is either directly forward or backward according as $\bar{\mu}_{IJ}$ is positive or negative. In short, we make the approximation that

$$W_i k_{ij}(\mu_0) \doteq a_{ij}\left[\tfrac{1}{2}(1 - |\bar{\mu}_{ij}|) + |\bar{\mu}_{ij}|\, \delta\left(\mu_0 - \frac{\bar{\mu}_{ij}}{|\bar{\mu}_{ij}|}\right) \right]$$

$$= W_i \tilde{k}_{ij}(\mu_0). \tag{5.2.11}$$

---

† Later, in Chapter 6, we make use of positive and negative *starting* weights: the total negative weight is usually small compared to the positive weight. If the negative weight is small and does not fluctuate from collision to collision, the effect on the variance, while deleterious, is tolerable.

‡ We refer specifically to the use of P_l expansions in thermal flux calculations. In fast calculations one usually takes P_l expansions in the center of mass and such expansions are not so troublesome. Note that it is not *impossible* to use P_l expansions in the thermal range, and many computer programs do so.

Clearly,

$$W_i \int_{-1}^{1} \tilde{k}_{ij}(\mu_0) \, d\mu_0 = a_{ij} = W_i \int_{-1}^{1} k_{ij}(\mu_0) \, d\mu_0,$$

$$W_i \int_{-1}^{1} \mu_0 \tilde{k}_{ij}(\mu_0) \, d\mu_0 = \bar{\mu}_{ij} a_{ij} = W_i \int_{-1}^{1} \mu_0 k_{ij}(\mu_0) \, d\mu_0.$$

Thus, the true and approximate scattering kernels have the same zeroth and first Legendre components. In this sense the above approximation resembles a simple P_1-approximation. However, in a P_1-approximation all Legendre components beyond the first vanish, while the higher components of $\tilde{k}_{ij}(\mu_0)$ all equal $\bar{\mu}_{ij} a_{ij} / W_i$.

When we try to evaluate approximations made in transport calculations, we find ourselves too often in the same unsatisfactory position. We do not have enough information to come to precise conclusions. Again we are in this position. We do not know much about the nature of errors caused by the above treatment of anisotropic scattering. There is some evidence that such errors are *usually* small. In a one-energy problem forward scattering is equivalent to no scattering. Therefore, if $\bar{\mu} \geq 0$, and in one energy, the treatment just described is a conventional transport approximation. Experience has shown that the transport approximation is often reasonably accurate. We may hope that the related approximation of Eq. (5.2.11) is also reasonably accurate. Nevertheless, it seems advisable to make a finer approximation whenever possible, and we shall discuss a refined anisotropic scattering kernel in Section 5.4.

5.3 THE MONTE CARLO CALCULATION

The discussion of the previous section has led us to a set of coupled transport equations

$$\boldsymbol{\omega} \cdot \nabla \psi_i(\mathbf{r}, \boldsymbol{\omega}) + \tilde{\Sigma}_{ti}(\mathbf{r}) \psi_i(\mathbf{r}, \boldsymbol{\omega}) = N_H(\mathbf{r}) \sum_{j=1}^{N} \int a_{ij}(\boldsymbol{\omega}' \cdot \boldsymbol{\omega}) \psi_j(\mathbf{r}, \boldsymbol{\omega}') \, d\boldsymbol{\omega}'$$

$$+ \frac{\Sigma_{sh}(\mathbf{r}) g_i(\mathbf{r})}{4\pi} + \frac{B_i}{4\pi} q(\mathbf{r}), \qquad 1 \leq i \leq N, \quad (5.3.1)$$

where $a_{ij}(\boldsymbol{\omega}' \cdot \boldsymbol{\omega}) = W_i k_{ij}(\boldsymbol{\omega}' \cdot \boldsymbol{\omega})$. To solve Eqs. (5.3.1), we require no techniques which have not already been discussed. In Chapter 4 we discussed methods for treating one-energy problems and in Chapter 2 we developed Monte Carlo methods for the solution of matrix equations. These methods may be combined in the formulation of an analog random walk process, and this point of view was taken in creating the MARC program. Proceeding along these lines neutrons are introduced isotropically into the system with source energies determined by the B_i and with spatial density $q(\mathbf{r})$, assumed constant over subregions. Scattering by heavy elements is assumed isotropic according to the transport approximation and does not alter the energy group of the scattered neutron. Scattering by hydrogen may result either in transfer from group j to group i or in a new direction in group j. The angular distribution of such a scattering (determined by Eq. 5.2.11) has a component which is isotropic and a component which is either forward or backward, depending on the sign of the average cosine in the group. Neutrons are scattered from group to group and region to region until they are ultimately absorbed in some group and region.

The basic random variable used to obtain estimates of any desired reaction rate is the modified track length estimator—estimator II of Table 2.1 (p. 80). This is combined with the capture estimator in such a way that the variance in each region is theoretically minimized. Use is made of Halperin's analysis (Ref. 6), as sketched in Section 3.11, to accomplish this minimization and obtain confidence intervals for the resulting random variable. Estimators are recorded for each energy group and region, and appropriate reaction rates are then obtained by summing over all groups and over edit regions, as required by the edit.

We have stressed repeatedly that it is often advantageous to solve the adjoint transport equation in place of the transport equation itself. This is true in multigroup as well as one-group problems. In multigroup, as in one-group calculations, the adjoint method is particularly well suited to the computation of flux averages over small regions, or fluxes at points. To formulate the multigroup adjoint method, we proceed very much as in Section 4.10.

Define vectors

$$\psi(\mathbf{r}, \boldsymbol{\omega}) = \begin{bmatrix} \psi_1(\mathbf{r}, \boldsymbol{\omega}) \\ \psi_2(\mathbf{r}, \boldsymbol{\omega}) \\ \cdot \\ \cdot \\ \cdot \\ \psi_N(\mathbf{r}, \boldsymbol{\omega}) \end{bmatrix}, \qquad g(\mathbf{r}) = \begin{bmatrix} g_1(\mathbf{r}) \\ g_2(\mathbf{r}) \\ \cdot \\ \cdot \\ \cdot \\ g_N(\mathbf{r}) \end{bmatrix}. \qquad (5.3.2)$$

In matrix notation Eq. (5.3.1) takes the form

$$\boldsymbol{\omega} \cdot \nabla \psi(\mathbf{r}, \boldsymbol{\omega}) + \tilde{\Sigma}_t(\mathbf{r})\psi(\mathbf{r}, \boldsymbol{\omega}) = N_H(\mathbf{r}) \int a(\boldsymbol{\omega} \cdot \boldsymbol{\omega}')\psi(\mathbf{r}, \boldsymbol{\omega}')\, d\boldsymbol{\omega}'$$
$$+ \frac{\Sigma_{sh}(\mathbf{r})}{4\pi} g(\mathbf{r}) + \frac{q(\mathbf{r})}{4\pi} B, \qquad (5.3.3)$$

where

$$g(\mathbf{r}) \equiv \int d\boldsymbol{\omega}\psi(\mathbf{r}, \boldsymbol{\omega}),$$

$$a \equiv (a_{ij}),$$

and B is the vector

$$B = (B_1, B_2, \ldots, B_N).$$

Again, as in Section 4.10, we write the transport equation in abbreviated form,

$$T\psi(\mathbf{r}, \boldsymbol{\omega}) = \frac{q(\mathbf{r})}{4\pi} B = Q(\mathbf{r}, \boldsymbol{\omega}) \equiv \frac{1}{4\pi} Q(\mathbf{r}),$$

and now we find

$$\int_V \phi^T(\mathbf{r}, \boldsymbol{\omega})T\psi(\mathbf{r}, \boldsymbol{\omega}) \, d\boldsymbol{\omega} \, d\mathbf{r} - \int_V \psi^T(\mathbf{r}, \boldsymbol{\omega})T^*\phi(\mathbf{r}, \boldsymbol{\omega}) \, d\boldsymbol{\omega} \, d\mathbf{r}$$
$$= \int_c \boldsymbol{\omega} \cdot \mathbf{n}\psi^T(\mathbf{r}, \boldsymbol{\omega})\phi(\mathbf{r}, \boldsymbol{\omega}) \, d\boldsymbol{\omega} \, d\mathbf{r}.$$

Here $\phi^T(\mathbf{r}, \boldsymbol{\omega})$ and $\psi^T(\mathbf{r}, \boldsymbol{\omega})$ are, respectively, transposes of the vectors ϕ and ψ, while

$$T^* \equiv \left[-\boldsymbol{\omega} \cdot \boldsymbol{\nabla} + \tilde{\Sigma}_t(\mathbf{r}) - \frac{\Sigma_{sh}(\mathbf{r})}{4\pi} \int d\boldsymbol{\omega}' - N_H(\mathbf{r}) \int d\boldsymbol{\omega}' a^T(\boldsymbol{\omega} \cdot \boldsymbol{\omega}') \right]. \quad (5.3.4)$$

Again, if ϕ and ψ obey reflecting boundary conditions,

$$\int_V \phi^T(\mathbf{r}, \boldsymbol{\omega})T\psi(\mathbf{r}, \boldsymbol{\omega}) \, d\boldsymbol{\omega} \, d\mathbf{r} = \int_V \psi^T(\mathbf{r}, \boldsymbol{\omega})T^*\phi(\mathbf{r}, \boldsymbol{\omega}) \, d\boldsymbol{\omega} \, d\mathbf{r}.$$

Let†

$$T^*f^*(\mathbf{r}, \boldsymbol{\omega}) = Q^*(\mathbf{r}, \boldsymbol{\omega}).$$

It follows that

$$\int_V [f^*(\mathbf{r}, \boldsymbol{\omega})]^T Q(\mathbf{r}, \boldsymbol{\omega}) \, d\boldsymbol{\omega} \, d\mathbf{r} = \int_V [\psi(\mathbf{r}, \boldsymbol{\omega})]^T Q^*(\mathbf{r}, \boldsymbol{\omega}) \, d\boldsymbol{\omega} \, d\mathbf{r}. \quad (5.3.5)$$

To compute a reaction rate in some region, R, we let‡

$$Q_i^*(\mathbf{r}, \boldsymbol{\omega}) \equiv \begin{cases} \frac{1}{4\pi}\Sigma_i(\mathbf{r}), & \mathbf{r} \in R, \\ 0, & \mathbf{r} \notin R. \end{cases}$$

If

$$g^*(\mathbf{r}) \equiv \int d\boldsymbol{\omega}f^*(\mathbf{r}, \boldsymbol{\omega}), \qquad \Sigma(\mathbf{r}) \equiv \text{diag}\left(\Sigma_1(\mathbf{r}), \Sigma_2(\mathbf{r}), \Sigma_N(\mathbf{r})\right),$$

then

$$\int_R g^T(\mathbf{r})\Sigma(\mathbf{r}) \, d\mathbf{r} = \int_V [g^*(\mathbf{r})]^T Q(\mathbf{r}) \, d\mathbf{r}.$$

Note that

$$\int_R g^T(\mathbf{r})\Sigma(\mathbf{r}) \, d\mathbf{r} \equiv \Sigma_{i=1}^N W_i \int_R \phi_i(\mathbf{r})\Sigma_i(\mathbf{r}) \, d\mathbf{r}$$

† We remark again that $f^*(\mathbf{r}, \boldsymbol{\omega})$ is the same as the $\chi^*(\mathbf{r}, \boldsymbol{\omega})$ (Eq. 3.7.26) of Chapter 3.
‡ Recall that the Δ_i are energy mesh widths, as on p. 179.

and that the sum on the right-hand side of the last equation can be interpreted as a trapezoidal integral over energy. Therefore,

$$\int_V g^T(\mathbf{r})\Sigma(\mathbf{r}) \, d\mathbf{r} \doteq \int_0^{E_o} \int_R dE \, d\mathbf{r}\phi(E,\mathbf{r})\Sigma(E,\mathbf{r}).$$

We see again, as in Chapter 4, that one may estimate reaction rates directly by solving the transport equation (i.e. by a "direct mode" calculation) or by solving the adjoint equations (via an "adjoint mode" calculation). Further, if we let the region R shrink to a point, we can, through use of the last equation, compute a reaction rate at that point.

We have noted, in Chapter 4, that the adjoint and transport equations for one-energy problems can both be simulated by the same Monte Carlo techniques. Unfortunately, we find, when we set out to solve the *multigroup* adjoint equations, that we face some new problems. Naturally we can, again, let

$$\psi^*(\mathbf{r}, -\mathbf{\omega}) \equiv f^*(\mathbf{r}, \mathbf{\omega}), \qquad g^*(\mathbf{r}) \equiv \int d\mathbf{\omega} f^*(\mathbf{r}, \mathbf{\omega}) = \int d\mathbf{\omega} \psi^*(\mathbf{r}, \mathbf{\omega}),$$

and it is clear that

$$\mathbf{\omega} \cdot \nabla \psi^*(\mathbf{r}, \mathbf{\omega}) + \tilde{\Sigma}_t(\mathbf{r})\psi^*(\mathbf{r}, \mathbf{\omega}) = N_H(\mathbf{r}) \int a^T(\mathbf{\omega} \cdot \mathbf{\omega}')\psi^*(\mathbf{r}, \mathbf{\omega}') \, d\mathbf{\omega}'$$

$$+ \frac{\Sigma_{sh}}{4\pi} g^*(\mathbf{r}) + \frac{1}{4\pi} Q^*(\mathbf{r}). \qquad (5.3.6)$$

The reader will observe that Eqs. (5.3.6) and (5.3.3) are quite similar in form, yet there is a difference between these two equations which, from our point of view, is very important. We see this most easily if we write the adjoint equations for an infinite homogeneous medium and examine the corresponding random walk process. In an infinite homogeneous medium

$$(\Sigma_a + \tilde{\Sigma}_{sH})g^* = N_H a^T g^* + Q^*, \qquad a^T = \int a^T(\mathbf{\omega}' \cdot \mathbf{\omega}) \, d\mathbf{\omega}'. \quad (5.3.7)$$

Expanding our notation, we write

$$(\Sigma_a + \tilde{\Sigma}_{sHi})g_i^* = N_H \sum_j a_{ij}^T g_j^* + Q_i^* = N_H \sum_j a_{ji} g_j^* + Q_i^*. \quad (5.3.8)$$

It seems natural, in simulating Eq. (5.3.8), to interpret $\tilde{\Sigma}_{sHi} g_i^*$ as a scattering rate at the i-th cut point (i.e. in the i-th group), and to treat the quantity $N_H a_{ji} g_j^*$ as the rate at which particles, emerging from scattering collisions at j, are transferred to i. Now,[†]

$$N_H \sum_i a_{ij} g_j = N_H \sum_i W_i \int_{-1}^1 \tilde{k}_{ij}(\mu_0) \, d\mu_0 g_j = \tilde{\Sigma}_{sHj} g_j,$$

[†] Note that, by definition (see p. 180) $\tilde{\sigma}_{sHi} \equiv \sum_{j=1}^N a_{ji}.$

but, generally,

$$N_H \sum_i a_{ij}^T g_j^* = N_H \sum_i a_{ji} g_j^* \neq \tilde{\Sigma}_{sHj} g_j^*. \tag{5.3.9}$$

The above inequality implies that, in the adjoint mode, the number of particles emerging from hydrogen scattering collisions is not equal to the number of hydrogen scattering collisions. In fact, when hydrogen scattering collision occurs in the i-th group, the number of emerging particles, or the net weight of all emerging particles, is equal to $\sum_{j=1}^N a_{ij}/\tilde{\sigma}_{sHi}$.

To make our argument somewhat more formal, we re-examine Eq. (5.3.6) and note that, if C_i is the hydrogen scattering source into the i-th group, then

$$C_i = N_H(\mathbf{r}) \sum_{j=1}^N \int a_{ij}^T(\boldsymbol{\omega}, \boldsymbol{\omega}') \psi_j^*(\mathbf{r}, \boldsymbol{\omega}') \, d\omega'$$

$$= N_H(\mathbf{r}) \sum_{j=1}^N \int a_{ji}(\boldsymbol{\omega} \cdot \boldsymbol{\omega}') \psi_j^*(\mathbf{r}, \boldsymbol{\omega}') \, d\omega'. \tag{5.3.10}$$

Defining $f_{ij}(\boldsymbol{\omega} \cdot \boldsymbol{\omega}') = a_{ij}(\boldsymbol{\omega} \cdot \boldsymbol{\omega}')/a_{ij}$, we may write

$$C_i = \sum_{j=1}^N \int d\omega' [N_H(\mathbf{r})\tilde{\sigma}_{sHj}\psi_j^*(\mathbf{r}, \boldsymbol{\omega}')] \left[\frac{a_{ji}}{\sum_{k=1}^N a_{jk}} \right] [f_{ji}(\boldsymbol{\omega} \cdot \boldsymbol{\omega}')] \left[\frac{\sum_{k=1}^N a_{jk}}{\tilde{\sigma}_{sHj}} \right]. \tag{5.3.11}$$

It will be seen that the first factor on the right-hand side of the above equation is the hydrogen scattering rate in group j. The second factor is the probability (in the adjoint mode) of scattering from j to i, while the third is the conditional probability density that, in a transition from j to i, the cosine of the scattering angle will be equal to $\boldsymbol{\omega} \cdot \boldsymbol{\omega}'$. But C_i is the net rate at which scattered particles are fed into the i-th group. Therefore, we must conclude that $K_j \equiv \sum_{k=1}^N a_{jk}/\tilde{\sigma}_{sHj}$ is the average number of scattered particles emerging from a single hydrogen scattering event in group j.†

Of course, there are many ways to incorporate the factor K_j into a random walk. One might actually use the K_j to determine the number of secondaries which emerge from hydrogen collisions. If $K_j < 1$, some hydrogen scattering collisions in group j will act like absorptions, and such artificial absorptions cause no real difficulties. On the other hand, if $K_j > 1$, hydrogen scattering in the j-th group will be a multiplicative process. If the K_j are substantially greater than one in many groups (as they generally will be), then the various starters may produce wildly fluctuating numbers of secondaries in the course of their histories. To follow the secondaries produced by multiplication is

---

† Note that in analyzing the direct mode scattering integral we would be led to define $K_j \equiv \sum_{k=1}^N a_{kj}/\tilde{\sigma}_{sHj} \equiv 1$. Thus, in the direct mode the average number of scattered particles per scattering event is one.

quite inconvenient in any Monte Carlo computation. Further, and this is much more serious, the fluctuation in the number of secondaries tends to raise the variance of all our Monte Carlo estimates.

However, we have another alternative. We may force the number of emergent particles per scattering collision to be equal to one, introducing the K_j as weights. Unfortunately, the K_j are not related in any simple way to the importance of scattering events in the adjoint mode. The use of nonunit weights which are not derived from a good importance function is as dangerous, in Monte Carlo, as the explicit use of a poor importance function. It is not surprising, then, to find (as we have) that adjoint calculations incorporating the K_j as weights are very inefficient.

Fig. 5.1

If we are to discuss the efficiency of the adjoint mode in more meaningful terms, we shall have to introduce efficiency criteria of some sort. We feel that it is useful to take, as one criterion, the variance in the following simple adjoint calculation. Given the problem configuration depicted in Fig. 5.1, we are to compute the absorption rate, A, in region II. Here regions I and II are identical in dimensions and composition, and the neutron source is uniform over region I. All exterior boundaries (shown as dashed lines) in Fig. 5.1 are cell boundaries. In a direct Monte Carlo computation we start histories in region I and make estimates in region II. On the other hand, in the adjoint mode we start histories in II and estimate in I. Now, since both regions are identical, it makes no difference whether we start histories in I and estimate in II or vice versa. Yet we find that, if the K_j are used as weights in the adjoint mode, the adjoint-mode variance in A is very much higher than the variance in a direct-mode computation. This difference in variances, since it is not due to geometric factors, must be caused by some inherent feature of the adjoint method, and our experience indicates that it is largely due to fluctuations in particle weights.

There is, in addition, a second criterion which we have found to be helpful. To judge the efficiency of a *multigroup* adjoint computation, we compare the variance in a *multigroup* activation rate with the variance in the same quantity computed, via the adjoint mode, in one group. Admittedly, it is difficult to construct a one-group problem which corresponds exactly to some multigroup problem, but it is easy (Ref. 7) to define a *roughly* equivalent one-group problem. The formulation of such one-group thermal problems is an art as old as reactor technology, an art which we cannot discuss here. It is noteworthy, however, that variances in multigroup adjoint calculations are much higher (if the K_j are used as weights) than in corresponding one-group adjoint calculations. Again our experience indicates that the large differences in variances are due to fluctuations in weights. It is clear that we must eliminate these fluctuations, and we show next how this may be done.

If we multiply Eq. (5.3.6) by the matrix ψ^∞, $\psi^\infty = \text{diag}(\psi_1^\infty, \psi_2^\infty, \dots, \psi_N^\infty)$, we find that

$$\boldsymbol{\omega} \cdot \nabla \eta_i(\mathbf{r}, \boldsymbol{\omega}) + \tilde{\Sigma}_{ti}(\mathbf{r})\eta_i(\mathbf{r}, \boldsymbol{\omega}) = N_H(\mathbf{r}) \sum_{j=1}^{N} \int d\boldsymbol{\omega}' b_{ji} f_{ji}(\boldsymbol{\omega} \cdot \boldsymbol{\omega}')\eta_j(\mathbf{r}, \boldsymbol{\omega}')$$
$$+ \frac{1}{4\pi}\Sigma_{sh}(\mathbf{r})\xi_i(\mathbf{r}) + \frac{1}{4\pi}\psi_i^\infty Q_i^*(\mathbf{r}), \quad (5.3.12)$$

where

$$\eta_i(\mathbf{r}, \boldsymbol{\omega}) \equiv \psi_i^\infty \psi_i^*(\mathbf{r}, \boldsymbol{\omega}), \qquad \xi_i(\mathbf{r}) \equiv \psi_i^\infty g_i^*(\mathbf{r}), \qquad b_{ji} \equiv \psi_i^\infty a_{ji}/\psi_j^\infty. \quad (5.3.13)$$

Equations (5.3.12) constitute a new set of multigroup equations which we shall solve, in place of (5.3.6), by Monte Carlo. At this point the ψ_i^∞ are, simply, undefined constants at our disposal. Now suppose that, in situations of interest to us, hydrogen occurs only in water. We may then write

$$\tilde{\Sigma}_{ti}(\mathbf{r}) = \Sigma_{aWi}(\mathbf{r}) + \Sigma_{ai}'(\mathbf{r}) + \Sigma_{sh}(\mathbf{r}) + \tilde{\Sigma}_{sHi}(\mathbf{r}).$$

In the above equation $\Sigma_{aWi}(\mathbf{r})$ is the absorption cross section of water, while $\Sigma_{ai}'(\mathbf{r}) \equiv \Sigma_{ai}(\mathbf{r}) - \Sigma_{aWi}(\mathbf{r})$. It will be seen that $\tilde{\Sigma}_{sHi}(\mathbf{r})$ and $\Sigma_{aWi}(\mathbf{r})$ do not occur explicitly in Eq. (5.3.12). Therefore, $\eta_i(\mathbf{r}, \boldsymbol{\omega})$ will be unaffected if we change $\tilde{\Sigma}_{sHi}(\mathbf{r})$ and $\Sigma_{aWi}(\mathbf{r})$ without changing their sum. Let us define

$$(\sigma_{sHi})_{\text{eff}} = \sum_{j=1}^{N} b_{ij} = \sum_{j=1}^{N} \frac{\psi_j^\infty}{\psi_i^\infty} a_{ij},$$

$$\sigma_{Wi} = [\tilde{\Sigma}_{sHi}(\mathbf{r}) + \Sigma_{aWi}(\mathbf{r})]/N_H(\mathbf{r}), \quad (5.3.14)$$

and

$$(\sigma_{aWi})_{\text{eff}} = \sigma_{Wi} - (\sigma_{sHi})_{\text{eff}}. \quad (5.3.15)$$

From (5.3.14) and (5.3.15) it follows that

$$\sigma_{Wi}\psi_i^\infty = \sum_{j=1}^{N} a_{ij}\psi_j^\infty + (\sigma_{aWi})_{\text{eff}}\psi_i^\infty$$
$$= \sum_{j=1}^{N} a_{ij}\psi_j^\infty + T_i, \quad (5.3.16)$$

and it follows from (5.3.14) that a single scattered particle, with unit weight, emerges from each hydrogen scattering collision. We are still free, in Eq. (5.3.16), to choose a set of values for the T's, and in choosing we shall be guided by the dual importance sampling theory developed in Chapter 3.

Let us take†

$$T_i \equiv \chi(E_i)W_i \Big/ \sum_{j=1}^{N} \chi(E_j)W_j. \qquad (5.3.17)$$

With T_i, so defined, Eqs. (5.3.16) may be regarded as thermal multigroup equations. In fact, they are multigroup equations for the flux in an infinite medium of water containing the thermal neutron source

$$Q(\mathbf{r}, E, \boldsymbol{\omega}) = \frac{1}{4\pi} \chi(E)N_H(\mathbf{r}).$$

Correspondingly, the ψ_i^∞ are infinite medium fluxes in water. Now, the fluxes ψ_i^∞ are, precisely, the importance functions appropriate to an infinite medium adjoint calculation. Therefore, in introducing the transformation (5.3.13) we are in effect using importance sampling techniques. This use of importance sampling has consequences which we shall soon see. At this point, however, we pause to examine some sampling techniques which have been used to simulate Eq. (5.3.12).

Suppose that we wish to estimate the activation rate, I, of some cross section, Σ_i, exposed in region R:

$$I = \sum_{j=1}^{N} \Sigma_i W_i \int_R \phi_i(\mathbf{r}) \, d\mathbf{r}.$$

Then $Q_i^*(\mathbf{r}) = \Sigma_i$. As usual, we let $Q_i(\mathbf{r})$ designate the source into the i-th group of the multigroup transport equation. Further, we assume, as in the past, that $Q_i(\mathbf{r}) = q(\mathbf{r})T_i$. Clearly, then,

$$\begin{aligned} I &= \sum_{i=1}^{N} T_i \int_V q(\mathbf{r})g_i^*(\mathbf{r}) \, d\mathbf{r} \\ &= \sum_{i=1}^{N} \frac{T_i}{\psi_i^\infty} \int_V q(\mathbf{r})\xi_i(\mathbf{r}) \, d\mathbf{r}, \end{aligned} \qquad (5.3.18)$$

where $\xi_i(\mathbf{r})$ is defined by Eq. (5.3.13).

But we know from Eq. (5.3.16) that

$$\frac{T_i}{\psi_i^\infty} = (\sigma_{aWi})_{\text{eff}}$$

and therefore

$$I = \sum_{i=1}^{N} \int_V \frac{q(\mathbf{r})}{N_H(\mathbf{r})} [\Sigma_{aWi}(\mathbf{r})]_{\text{eff}} \xi_i(\mathbf{r}) \, d\mathbf{r}. \qquad (5.3.19)$$

In the MARC program it is assumed that $q(\mathbf{r})$ and $N_H(\mathbf{r})$ are piecewise

---

† The χ's and W's are defined on p. 179.

constant over subregions of V, i.e. $q(\mathbf{r}) = q_m$, $N_H(\mathbf{r}) = N_m$ if $\mathbf{r} \in R_m$. Thus, in MARC

$$I = \sum_m \frac{q_m}{N_m} \left[\sum_{i=1}^{N} (\Sigma_{aWi}^m)_{\text{eff}} \int_{R_m} \xi_i(\mathbf{r}) \, d\mathbf{r} \right]. \tag{5.3.20}$$

Now, in Eq. (5.3.12) $\xi_i(\mathbf{r})$ plays the role of a scalar flux, and the true macroscopic absorption cross section of water is replaced, in group i and region m, by $(\Sigma_{aWi}^m)_{\text{eff}}$. Therefore, the quantity in square brackets above is the effective absorption rate in the water contained within region R_m. We estimate this absorption rate just as we would any other absorption rate.

a) We choose a starting location and group from the density

$$p(\mathbf{r}, i) = \psi_i^\infty Q_i^* \Big/ V_R \sum_{j=1}^{N} \psi_j^\infty Q_j^*$$

$$= \psi_i^\infty \Sigma_i \Big/ V_R \sum_{j=1}^{N} \psi_j^\infty \Sigma_j, \qquad \mathbf{r} \in R,$$

$$p(\mathbf{r}, i) = 0, \qquad \mathbf{r} \notin R.$$

Here V_R is the volume of the region R.

b) We assign to each starter the weight $W \equiv V_R \sum_{j=1}^{N} \psi_j^\infty \Sigma_j$.

c) For each starter we construct a sample history governed by Eq. (5.3.12). If a starter is absorbed in R_m, in group i, we record the absorbed weight

$$(Wq_m/N_m)(\Sigma_{aWi}^m)_{\text{eff}}/(\Sigma_{ai}^m)_{\text{eff}}.$$

The quantity $(\Sigma_{ai}^m)_{\text{eff}}$ is the sum, in R_m and group i, of the effective absorption cross section of water and the true absorption cross section of all other materials. Finally, the average absorbed weight per starter is an unbiased estimate (an "absorption estimate") of I. We also compute, by familiar methods, track lengths and combined estimates of I.

In an infinite medium of pure water the ratio $(\Sigma_{aWi})_{\text{eff}}/(\Sigma_{ai})_{\text{eff}}$ is equal to one. Therefore, every starter contributes to the absorption estimator the absorbed weight Wq/N_H. One can show that this weight is indeed equal to the activation rate $I = V_R \sum_j \Sigma_j W_j \phi_j$, and it is clear that the absorption estimator is a zero-variance estimator of I. This is true precisely because we have smuggled importance sampling into the adjoint calculation. The ψ_i^∞, as we have noted earlier, are the importance functions appropriate to the infinite-medium adjoint calculation, and Eqs. (5.3.13) are precisely the transformations generated by importance sampling.

Consider, again, the problem configuration depicted in Fig. 5.1. Suppose that $Q_i(\mathbf{r})$, the source density in region I, is equal to qT_i. Then the absorption rate, A, in II is equal to $V_I q P_{\text{I} \to \text{II}}$, where $P_{\text{I} \to \text{II}}$ is the average probability that a starter will be absorbed in II. In a direct-mode calculation of A the relative variance, σ_r, of the absorption estimator will be equal to $P_{I \to II}/(1 - P_{I \to II})$. In the adjoint mode we find that

$$A = Wq P^*_{II \to I}/N_H,$$

where $P^*_{II \to I}$ is the average probability that an *adjoint-mode* starter will be absorbed in I. In the adjoint mode the relative variance, σ^*_r, of the absorption estimator is equal to $P^*_{II \to I}/(1 - P^*_{II \to I})$. Now,

$$\frac{Wq}{N_H} = V_{II} \frac{q}{N_H} \sum_{j=1}^{N} \Sigma_{aWj} \psi_j^{\infty},$$

and it can easily be shown, by summing Eqs. (5.3.16), that $\sum_{j=1}^{N} \Sigma_{aWj} \psi_j^{\infty} = N_H$. Therefore,

$$\frac{Wq}{N_H} = V_{II} q.$$

Since

$$\frac{Wq}{N_H} P^*_{II \to I} = V_I q P_{I \to II} \quad \text{and} \quad V_I = V_{II}$$

we can conclude that $\sigma_r = \sigma^*_r$. Thus, absorption estimators in the direct and adjoint modes have the same variance. Unfortunately, we cannot analyze the track length estimators theoretically, but numerical experiments show that these, too, have roughly the same variance. We find, thus, that the direct and adjoint calculations are equally efficient, and in the given problem configurations we have no right to expect more than this of the adjoint mode.

Above we have compared the performance of the direct and adjoint modes in a fairly simple situation. It is interesting, also, to compare the efficiency of the adjoint mode in a multigroup computation with its efficiency in a corresponding one-group computation. Unfortunately, a comparison of variances in the adjoint mode leads to no rigorous and precise conclusion. One can show, however, that if, in a given problem configuration,

a) all neutron sources are confined to regions containing pure water, and

b) the spectrum everywhere is roughly the same as the infinite medium water spectrum,

then adjoint activation estimates have roughly the same variance in multigroup and corresponding one-group calculations.

5.4 REFINEMENTS OF THE FORMULATION

In Sections 5.2 and 5.3 we have discussed one practical approach to the Monte Carlo treatment of thermal neutrons, but, of course, our formulation of the thermal multigroup equations is not the only possible formulation. The accuracy of the multigroup equations can be improved in many ways.

It is true that for each improvement we must be prepared to pay a price of some sort. On the other hand, as computers grow larger and faster, it is quite natural to demand more and more accuracy in our computational methods.

Perhaps one weakness of the multigroup equations we have developed is more obvious, more disturbing, than all others. It is clear that we have grossly distorted the angular distribution of neutrons scattered by hydrogen. As pointed out in Section 5.2, only the P_0 and P_1 components of the P_1-MARC scattering kernel are at our disposal. We may adjust these components as we like, but once the P_0 and P_1 components are fixed all other components are completely determined. Thus, we are forced, in the P_1-MARC, to fit the angular distribution of scattered neutrons (for given initial and final energies) with only two parameters.

It seems reasonable to expect that one would fit the true† angular distribution more closely with the aid of more parameters. Suppose, for example, that we are given k_{ijl} for $0 \leq l \leq 3$. In other words, we are given the first *four* Legendre moments of the true differential scattering kernel. We may concoct various approximate kernels, $\tilde{k}_{ij}(\mu_0, \alpha_1, \alpha_2, \alpha_3, \alpha_4)$, containing four parameters which can be adjusted so that

$$\int_{-1}^{1} P_l(\mu_0) \tilde{k}_{ij}(\mu_0, \alpha_{1ij}, \alpha_{2ij}, \alpha_{3ij}, \alpha_{4ij}) \, d\mu_0 \equiv \tilde{k}_{ijl} = k_{ijl}, \qquad l = 0, 1, 2, 3.$$
$$(5.4.1)$$

But, having fixed the parameters so as to satisfy these last equations we should like the approximate kernel to be nonnegative. Unfortunately, we see no way to guarantee that this will be true. More precisely, we see no way to construct a simple function, $\tilde{k}(\mu_0, \alpha_1, \alpha_2, \alpha_3, \alpha_4)$, which is necessarily nonnegative if its first four moments are equal to those of some given, nonnegative function.

Facing an impasse, we feel we must retreat. We shall not attempt to match the full-range moments of the true scattering kernel. Instead we match the first four *half-range* moments. Let

$$a_{ij}^+ \equiv W_i \int_0^1 k_{ij}(\mu_0) \, d\mu_0,$$

$$\mu_{ij}^+ a_{ij}^+ \equiv W_i \int_0^1 \mu_0 k_{ij}(\mu_0) \, d\mu_0,$$

$$a_{ij}^- \equiv W_i \int_{-1}^0 k_{ij}(\mu_0) \, d\mu_0,$$

---

† By the "true" angular distribution we mean here the angular distribution given by some reasonable theory of molecular binding. Detailed experimental information on scattering kernels is accumulating rapidly, but empirical kernels are not as easy to use as computed kernels.

and

$$\mu_{ij}^- a_{ij}^- \equiv W_i \int_{-1}^0 |\mu_0|\, k_{ij}(\mu_0)\, d\mu_0.$$

If $\mu_{ij}^+ \geq \frac{1}{2}$, define

$$W_i \tilde{k}_{ij}(\mu_0) = a_{ij}^+[2(1 - \mu_{ij}^+) + (2\mu_{ij}^+ - 1)\delta(\mu_0 - 1)], \qquad \mu_0 \geq 0. \quad (5.4.2)$$

Otherwise

$$W_i \tilde{k}_{ij} \equiv a_{ij}^+[2\mu_{ij}^+ + (1 - 2\mu_{ij}^+)\delta(\mu_0)], \qquad \mu_0 \geq 0. \quad (5.4.3)$$

It is easy to show that

$$W_i \int_0^1 \tilde{k}_{ij}(\mu_0)\, d\mu_0 = a_{ij}^+ = W_i \int_0^1 k_{ij}(\mu_0)\, d\mu_0,$$

$$W_i \int_0^1 \mu_0 \tilde{k}_{ij}(\mu_0)\, d\mu_0 = a_{ij}^+ \mu_{ij}^+ = W_i \int_0^1 \mu_0 k_{ij}(\mu_0)\, d\mu_0.$$

Similarly, if $\mu_{ij}^- \geq \frac{1}{2}$,

$$W_i \tilde{k}_{ij} \equiv a_{ij}^-[2(1 - \mu_{ij}^-) + (2\mu_{ij}^- - 1)\delta(\mu_0 + 1)], \qquad \mu_0 < 0, \quad (5.4.4)$$

and otherwise

$$W_i \tilde{k}_{ij} \equiv a_{ij}^-[2\mu_{ij}^- + (1 - 2\mu_{ij}^-)\delta(\mu_0)], \qquad \mu_0 < 0. \quad (5.4.5)$$

Again

$$W_i \int_{-1}^0 k_{ij}(\mu_0)\, d\mu_0 = W_i \int_{-1}^0 \tilde{k}_{ij}(\mu_0)\, d\mu_0,$$

$$W_i \int_{-1}^0 |\mu_0|\, k_{ij}(\mu_0)\, d\mu_0 = W_i \int_{-1}^0 |\mu_0|\, \tilde{k}_{ij}(\mu_0)\, d\mu_0.$$

Thus, Eqs. (5.4.2) through (5.4.5) define a kernel which is nonnegative and whose first four half-range moments match those of the true kernel. It is this kernel which is used in the latest version of MARC.

It should be noted that

$$a_{ij} = a_{ij}^+ + a_{ij}^-, \qquad a_{ij}\bar{\mu}_{ij} = a_{ij}^+\mu_{ij}^+ - a_{ij}^-\mu_{ij}^-.$$

Therefore, any kernel whose first four half-range moments are correct certainly has correct P_0 and P_1 full-range moments. In matching the first four half-range moments we have also matched the first two full-range moments.

Perhaps it will occur to the reader that a very much simpler treatment of anisotropic scattering is perfectly feasible. We refer here to a method which might be called the "histogram method" because the differential scattering kernel is represented as a histogram. In one version of the histogram method we divide the interval $-1 \leq \mu_0 \leq 1$ into M equal subintervals. Let $\mu_0^{(n)}$ and

$\mu_0^{(n+1)}$ be, respectively, the left-hand and right-hand boundaries of the n-th subinterval, while

$$a_{ij}\Pi_{ij}^{(n)} = W_i \int_{\mu_0^{(n)}}^{\mu_0^{(n+1)}} k_{ij}(\mu_0), \qquad n = 1, 2, \ldots, M.$$

As before,

$$a_{ij} \equiv W_i \int_{-1}^{1} k_{ij}(\mu_0) \, d\mu_0$$

so that

$$\sum_{n=1}^{M} \Pi_{ij}^{(n)} = 1.$$

Define

$$W_i \check{k}_{ij}(\mu_0) \equiv a_{ij}\Pi_{ij}^{(n)}, \qquad \mu_0^{(n)} < \mu_0 \leq \mu_0^{(n+1)}, \qquad n = 1, 2, \ldots, M.$$

It will be seen that $\check{k}_{ij}(\mu_0)$ is constant over each subinterval and that $\Pi_{ij}^{(n)}$ is precisely the probability that a scattered neutron, in going from E_j to E_i, will enter the n-th subinterval.

Actually, there seems to be no good reason why all the subintervals should be equal in size. Probably it is better to make all the *probabilities*, $\Pi_{ij}^{(n)}$, equal in size. The endpoints of the angular interval would then be determined, implicitly, by the equations

$$\frac{W_i}{a_{ij}} \int_{\mu_0^{(n)}}^{\mu_0^{(n+1)}} k_{ij}(\mu_0) \, d\mu_0 = \frac{1}{M}, \qquad n = 1, 2, \ldots, M. \qquad (5.4.6)$$

Clearly, the angular mesh defined via Eq. (5.4.6) will tend to be densest in angular regions where the differential scattering kernel is largest.

It is obvious that in the histogram method M is perfectly general. On the other hand, one cannot easily alter the number of degrees of freedom incorporated into the MARC kernel. To introduce two extra degrees of freedom into the angular distribution it was necessary, as we have seen, to make basic changes in the structure of the MARC kernel. What advantage, then, has the MARC method over the histogram method? We shall show that the method used in MARC has two advantages. First of all, computations with this histogram kernel are relatively slow. To see that this is true we examine, in detail, the calculations of the final direction vector, $\mathbf{\Omega}'$.

Suppose that we are given the initial energy, E_j, and the final energy E_i. Let $P(n)$ be the probability that μ_0 will lie in the n-th angular subinterval. We already know that $P(n) = \Pi_{ij}^{(n)}$, and we can draw an n from $P(n)$ by the usual method.[†] Given n we take μ_0 from a distribution uniform in the n-th subinterval. We now have determined the scalar product $\mu_0 \equiv \mathbf{\omega} \cdot \mathbf{\omega}'$, but we have yet to fix that component of $\mathbf{\omega}'$ which is perpendicular to $\mathbf{\omega}$. At this point

† See Section 1.5.

it is convenient to introduce a coordinate system fixed to $\boldsymbol{\omega}$, in addition to the conventional (x, y, z)-system fixed in the laboratory. Let $\mathbf{i}$, $\mathbf{j}$, and $\mathbf{k}$, respectively, be unit vectors along the x-, y-, and z-axis. Let $\mathbf{i}' \equiv \boldsymbol{\omega}$ and

$$\mathbf{k}' \equiv [\mathbf{k} - (\boldsymbol{\omega} \cdot \mathbf{k})\boldsymbol{\omega}]/[1 - (\boldsymbol{\omega} \cdot \mathbf{k})^2]^{1/2}$$

$$= [\mathbf{k} - \omega_z \boldsymbol{\omega}]/[1 - \omega_z^2]^{1/2}.$$

One can easily show that $\mathbf{k}'$ is a unit vector orthogonal to $\mathbf{i}'$. Finally, define

$$\mathbf{j}' = \mathbf{k}' \times \mathbf{i}' = \mathbf{k} \times \boldsymbol{\omega}/[1 - \omega_z^2]^{1/2}.$$

The reader may verify that $\mathbf{i}'$, $\mathbf{j}'$, and $\mathbf{k}'$ form an orthogonal, right-handed, system of base vectors so that $\mathbf{i}' \times \mathbf{j}' = \mathbf{k}'$, $\mathbf{j}' \times \mathbf{k}' = \mathbf{i}'$, $\mathbf{k}' \times \mathbf{i}' = \mathbf{j}'$. In terms of the primed coordinate vectors

$$\boldsymbol{\omega}' = \mu_0 \mathbf{i}' + \mathbf{j}'[1 - \mu_0^2]^{1/2} \cos \phi + \mathbf{k}'[1 - \mu_0^2]^{1/2} \sin \phi, \qquad (5.4.7)$$

where ϕ is an angle drawn from a uniform distribution on the interval $\{0, 2\pi\}$. Given Eqs. (5.4.7) one can compute the components of $\boldsymbol{\omega}'$ in the original coordinate system. Note first that

$$\mathbf{i} \cdot \mathbf{i}' = \omega_x, \qquad (5.4.8)$$

$$\mathbf{i} \cdot \mathbf{j}' = \mathbf{i} \cdot \mathbf{k} \times \boldsymbol{\omega}[1 - \omega_z^2]^{-1/2} = \mathbf{i} \times \mathbf{k} \cdot \boldsymbol{\omega}[1 - \omega_z^2]^{-1/2}$$

$$= -\omega_y[1 - \omega_z^2]^{-1/2}, \qquad (5.4.9)$$

$$\mathbf{i} \cdot \mathbf{k}' = -\omega_z \omega_x[1 - \omega_z^2]^{-1/2}. \qquad (5.4.10)$$

Similarly,

$$\mathbf{j} \cdot \mathbf{i}' = \omega_y, \qquad (5.4.11)$$

$$\mathbf{j} \cdot \mathbf{j}' = \omega_x[1 - \omega_z^2]^{-1/2}, \qquad (5.4.12)$$

$$\mathbf{j} \cdot \mathbf{k}' = -\omega_z \omega_y[1 - \omega_z^2]^{-1/2}, \qquad (5.4.13)$$

$$\mathbf{k} \cdot \mathbf{i}' = \omega_z, \qquad (5.4.14)$$

$$\mathbf{k} \cdot \mathbf{j}' = 0, \qquad (5.4.15)$$

$$\mathbf{k} \cdot \mathbf{k}' = [1 - \omega_z^2]^{1/2}. \qquad (5.4.16)$$

It follows that

$$\omega_x' = \mu_0 \omega_x - [1 - \mu_0^2]^{1/2}[\omega_y \cos \phi + \omega_z \omega_x \sin \phi][1 - \omega_z^2]^{-1/2}, \quad (5.4.17)$$

$$\omega_y' = \mu_0 \omega_y + [1 - \mu_0^2]^{1/2}[\omega_x \cos \phi - \omega_z \omega_y \sin \phi][1 - \omega_z^2]^{-1/2}, \quad (5.4.18)$$

and

$$\omega_z' = \mu_0 \omega_z + [1 - \mu_0^2]^{1/2}[1 - \omega_z^2]^{1/2} \sin \phi. \qquad (5.4.19)$$

Evidently, the computation of μ_0 and ϕ, and the subsequent evaluation of the components of $\boldsymbol{\omega}'$ (with the aid of the last three equations), involve a good deal of work.

In the P_1-MARC we deal only with forward, backward, or isotropic scattering. If the scattering is forward, $\omega' = \omega$: it is unnecessary to compute ω'. For backward scattering $\omega' = -\omega$ and the computation of ω' is trivial. When the scattering is isotropic, we form ω' exactly as in Chapter 1. The selection of a unit vector from an isotropic distribution is a fairly simple process.

Of course, it is somewhat more expensive to use the improved MARC kernel. The differential scattering kernel defined by Eqs. (5.4.2) through (5.4.5) may induce collisions with $\mu_0 = 0$. When such collisions occur, it is necessary, again, to compute an azimuthal angle, ϕ, and to determine the components of ω' from Eqs. (5.4.17), (5.4.18), and (5.4.19). If they do not occur too frequently, problem running times will not be seriously affected. We have found in practice that the use of the improved kernel in place of the original MARC kernel usually increases running times by less than 5%.

We believe (though we are not sure) that the MARC method has another advantage over the histogram method. It will be seen that the histogram kernel has correct P_0-moments:

$$W_i \int_{-1}^{1} \tilde{k}_{ij}(\mu_0)\,d\mu_0 = \sum_{n=1}^{N} a_{ij}\Pi_{ij}^{(n)} = a_{ij}$$

$$= W_i \int_{-1}^{1} k_{ij}(\mu_0)\,d\mu_0.$$

However, it does not necessarily have correct P_1-moments. In general,

$$\int_{-1}^{1} \tilde{k}_{ij}(\mu_0)\mu_0\,d\mu_0 \neq \int_{-1}^{1} k_{ij}(\mu_0)\mu_0\,d\mu_0,$$

and the errors in the P_1-moments may be quite large if N is as small as two or four. Now, in one energy group, Σ_{s1}, the P_1-moment of the scattering cross section enters directly into the transport cross section. It is known that the value of the transport cross section often has a strong influence on the flux shape. In fact, in the transport approximation *only* the transport and absorption cross sections influence the flux. We see, then, that the P_1-moment, Σ_{s1}, seems to be important. Therefore, we feel that the MARC kernel probably gives more accurate flux shapes than histogram kernels when the MARC and histogram kernels have the same number of free parameters.

We set out above to improve the angular distribution of scattered neutrons by introducing into the MARC scattering kernel four adjustable parameters. It is also possible to refine our treatment of the *energy* distribution of scattered neutrons. In deriving the MARC transfer matrices, we postulated, in effect, that neutrons could take on only certain discrete energy values. One could instead divide the thermal energy band into subintervals and allow the energy to take on any value in any subinterval. This approach has important advantages when there are resonances in the thermal band. If energy is regarded as a continuous variable, it is possible to represent the resonance cross sections with a very fine mesh, while using a transfer matrix appropriate to a relatively coarse mesh. The matrix elements, a_{ij}, then determine the probability that, in a moderating collision, a neutron will be transferred from the j-th subinterval to the i-th. Given that the neutron is to fall into the i-th subinterval, its final energy is drawn from a distribution uniform over that subinterval.

Finally, we note that it is not necessary to separate as sharply as we have the thermal and epithermal energy bands. One can cover both ranges with a single Monte Carlo calculation. In the RECAP (Ref. 8) and DRAM (Ref. 9) programs the thermal and epithermal ranges are coalesced. However, at this time neither program treats epithermal neutrons rigorously.

5.5 TREATMENT OF SMALL PERTURBATIONS

Frequently, in reactor design and analysis work, it is necessary to compute small changes in fluxes and activation rates, changes produced by perturbations in composition or geometry. Small perturbations may result, for example, from temperature changes, design modifications, or, perhaps, from the insertion of a detector foil into a reactor lattice. In principle it is possible to estimate perturbations by running independent perturbed and unperturbed Monte Carlo calculations. In practice, however, this straightforward approach is often unsatisfactory. Let us suppose that we are interested in computing the difference, Δ, between the numbers A and B, and that $A \doteq B$. Then, if A and B are computed independently,

$$\sigma_R^2(\Delta) \equiv \frac{\sigma^2(\Delta)}{\Delta^2} = \frac{\sigma^2(A) + \sigma^2(B)}{\Delta^2}$$

$$= \frac{A^2\sigma_R^2(A) + B^2\sigma_R^2(B)}{\Delta^2} \doteq \left(\frac{A}{\Delta}\right)^2 [\sigma_R^2(A) + \sigma_R^2(B)],$$

where $\sigma_R^2(A) \equiv \sigma^2(A)/A^2$, $\sigma_R^2(B) = \sigma^2(B)/B^2$. Thus, if $|A/\Delta| \gg 1$, the relative error in Δ will be much greater than the relative errors in A and B. Such amplification of errors can only be avoided through the use of correlated sampling techniques.

One method of correlation, already described in Section 3.9, is incorporated in MARC. In all MARC problems the first history starts with the same pseudo-random number, ξ_{11}. Similarly, second histories always start with the same number, ξ_{21}, etc. As a result, small changes in input tend to produce small and statistically significant changes in estimated fluxes and activations. If problem parameters are perturbed only slightly, many histories in the perturbed and unperturbed problems will be identical. These histories will contribute no variance to the difference between estimates of the perturbed and unperturbed reaction rates. Quite commonly one finds that this very simple correlation technique reduces variances in differences by factors of 10, or more.

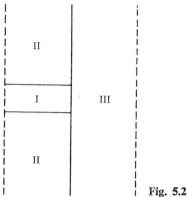

Fig. **5.2**

Nevertheless, much more effective correlation methods can be devised. In fact, as we have already seen (Section 3.9), estimates of perturbed and unperturbed reaction rates may be based on one single set of histories, so that the perturbed and unperturbed computations are completely correlated. To achieve complete correlation between perturbed and unperturbed problems is difficult if the perturbation is complicated. A program designed to handle *any* perturbation by this method of complete correlation would be very elaborate. However, it is easy to use complete correlation to treat perturbations of simple types. Suppose, for example, that we wish to study the perturbation produced by a foil (depicted in Fig. 5.2 as region I) placed in a fuel rod (region II). In Fig. 5.2 region III is filled with water, and only region III contains a source. For the sake of simplicity we deal with this problem, first, in one energy group. In the unperturbed situation the cross sections in region I are the same as the fuel cross sections, while the perturbed cross sections are, of course, the cross sections of the foil. We shall assume that scattering in the foil and fuel are isotropic.† To compute the average flux in region I, we use the adjoint mode; that is, we put a unit source in region I and estimate the integral of the adjoint flux in II. By reciprocity

$$\bar{\phi}_{\mathrm{I}} A_{\mathrm{I}} = Q_{\mathrm{II}} \int_{\mathrm{II}} \phi^*(\mathbf{r}) \, d\mathbf{r}. \qquad (5.5.1)$$

† This is not a severe restriction, since an anisotropic cross section can be replaced by an equivalent isotropic cross section via the transport approximation. The transport approximation should be adequate for the treatment of heavy elements.

With the aid of Eq. (5.5.1) we propose to compute the perturbed and un-perturbed fluxes simultaneously, and the computation will proceed as follows.

To estimate the unperturbed flux we put the source in I and run a Monte Carlo governed by the unperturbed transport equation. In generating histories we take the unperturbed cross sections to be the cross sections of region I, but we imagine that two particles are simultaneously participating in each event. One particle (the "unperturbed particle") has a weight of one. The other starts its history with a weight of one, but its weight is modified whenever it enters region I. Whenever it travels a distance l in this region, its current weight is multiplied by $\exp [(\Sigma_{tIU} - \Sigma_{tIP})l]$. When it scatters, its weight is multiplied by $\Sigma_{sIP}/\Sigma_{sIU}$.† Here Σ_{tIP} and Σ_{tIU} are, respectively, perturbed and unperturbed total cross sections in I. Correspondingly, Σ_{sIU} and Σ_{sIP} are unperturbed and perturbed scattering cross sections. In computing the integral of the adjoint flux, any convenient estimator may be used, but we must take note of the weights of the perturbed particles. Thus, if we use track lengths to estimate the unperturbed flux, we must use *weighted* track lengths to estimate the perturbed flux. If, to estimate the unperturbed flux, we count absorptions in region II, then the total weight absorbed in region II gives us an estimate of the perturbed flux, and so on.

The method just described is applicable to multigroup, as well as one-group problems, if the foil and fuel contain no hydrogen. Of course, it is necessary in multigroup calculations to store perturbed and unperturbed cross sections for each group. When a particle in group i enters region I, the group i cross sections must be used to modify its weight. If there is hydrogen in region I, or if the foil is placed in the water, the computational procedure must be modified slightly, but it remains quite feasible and simple.

Before closing this chapter we wish to discuss a problem which has interested reactor physicists (experimental physicists in particular) for several years. This is a particularly simple perturbation problem and it is interesting partly because it is so simple. A small foil is placed in an infinite water-bath. The water contains a uniform and isotropic slowing-down source and we are required to compute the activation of the foil.

We treat this problem, first, by the method of superposition. In the notation of Section 5.3 let the source density in the i-th group be T_i. Designate the foil as "region I" and call the water region "region II." Define a problem β such that

$$Q_{\beta i}(\mathbf{r}) = \begin{cases} \Sigma_{afi}\varphi_i^\infty/N_H, & \mathbf{r} \in R_{\mathrm{I}}, \\ 0, & \mathbf{r} \in R_{\mathrm{II}}, \end{cases} \quad i = 1, 2, \ldots, N.$$

---

† The procedure just described is unbiased if and only if there is no true source in I, i.e. if $Q_{\mathrm{I}} = 0$. However, a very similar procedure may be used if the foil is placed in the water, and not in the fuel. Note that Σ_{sIU} must not be zero unless $\Sigma_{sIP} = 0$. If this condition is violated, we must interchange the definitions of perturbed and unperturbed problems.

Here Σ_{afi} is the foil's absorption cross section in the i-th group. If A_f is the activation rate in the foil, one can show that

$$A_f = \sum_{j=1}^{N} \Sigma_{aWj} \int_{\text{II}} g_{\beta j}(\mathbf{r})\, d\mathbf{r}$$

$$= V_f \left[\sum_{j=1}^{N} \Sigma_{afj} \psi_j^{\infty} \right] P_{\text{I}\to\text{II}} / N_H.$$

But $V_f \sum_{j=1}^{N} \Sigma_{afj} \psi_j^{\infty} / N_H$ is precisely equal to the activation, A_{fu}, of the foil in the absence of any flux perturbation. Therefore,

$$R \equiv \frac{A_f}{A_{fu}} = P_{\text{I}\to\text{II}}. \tag{5.5.2}$$

Let $P_{0(\text{I})}$ be the average probability that a starter, drawn from the problem β source, will enter the water. Let Π be the conditional probability that a starter, having entered the water, will be absorbed there. Then

$$P_{\text{I}\to\text{II}} = P_{0(\text{I})} \Pi. \tag{5.5.3}$$

From Eqs. (5.5.2) and (5.5.3) we see that there are two ways to estimate R. First, we may use Monte Carlo in problem β to calculate $P_{\text{I}\to\text{II}}$ directly. Second, we may calculate $P_{0(\text{I})}$ independently, using Monte Carlo in problem β solely to determine Π. If probabilities are estimated binomially, then, in the first case, the relative variance in R is equal to $(1 - P_{\text{I}\to\text{II}})/P_{\text{I}\to\text{II}}$, and, in the second, it is equal to $(1 - \Pi)/\Pi$. It is easy to see that the second variance will always be smaller than the first.

As for the escape probability, $P_{0(\text{I})}$, this quantity, too, may be computed in various ways. If, in problem β, we replace the water by a pure absorber, the absorption probability in region II is equal to $P_{0(\text{I})}$. Further, if we eliminate all scattering in region II, the Monte Carlo histories are likely to be very short, so that this computation of $P_{0(\text{I})}$ will be efficient and inexpensive.

In the adjoint mode we put into region I the source density

$$Q_i^*(\mathbf{r}) = \Sigma_{afi} \psi_i^{\infty}, \qquad i = 1, 2, \ldots, N$$

and compute

$$A_f = \frac{1}{N_H} \sum_{j=1}^{N} (\Sigma_{aWj})_{\text{eff}} \int_{\text{II}} g_j^*(\mathbf{r})\, d\mathbf{r}.$$

Again we find that

$$R \equiv \frac{A_f}{A_{fu}} = P_{\text{I}\to\text{II}}^*, \tag{5.5.4}$$

where $P_{\text{I}\to\text{II}}^*$ is the average probability that a starter in the adjoint mode will be absorbed in II. On comparing Eqs. (5.5.3) and (5.5.4), we see that $P_{\text{I}\to\text{II}}^* = P_{\text{I}\to\text{II}}$. Suppose, for simplicity, that all probabilities are estimated binomially.

Then the method of superposition and the reciprocity method are equally efficient in the sense that both give the same variance in R for the same number of histories. Just as before, we can, if we wish, separate $P^*_{\mathrm{I}\to\mathrm{II}}$ into two factors,

$$P^*_{\mathrm{I}\to\mathrm{II}} = P_{0(\mathrm{I})}\Pi,$$

and compute $P_{0(\mathrm{I})}$ separately.

We have seen in Chapter 4 that, in one energy group, there is a close relation between the adjoint method and the method of superposition. Clearly, this is also true in certain types of multigroup problems. It is not true in all multienergy problems and, in particular, it is not true in the resonance escape problem which we consider next.

REFERENCES

1. E. M. GELBARD, H. B. ONDIS, and J. SPANIER, "MARC—A Multigroup Monte Carlo Program for the Calculation of Capture Probabilities," *WAPD-TM*-273 (1962).

2. H. C. HONECK, *THERMOS—A Thermalization Transport Theory Code for Reactor Lattice Calculations*, Brookhaven National Laboratory (1961).

3. MARK NELKIN, "Scattering of Slow Neutrons by Water," *Phys. Rev.*, **119**, No. 2 (1960).

4. W. W. CLENDENIN, D. E. GEORGE, and B. S. HAMILL, "MAGMA—A Philco-2000 Program for the Calculation of Scattering Kernels, Neutron Spectra and Few Group Parameters for Thermal Neutrons," *WAPD-TM*-373 (September, 1964).

5. M. M. R. WILLIAMS, *The Slowing Down and Thermalization of Neutrons*, John Wiley and Sons, Inc., New York (1966).

6. M. HALPERIN, "Almost linearly-optimum combination of unbiased estimates," *J. Am. Stat. Assoc.*, **56**, 36 (1961).

7. *Naval Reactors Physics Handbook*, Vol. I, Selected Basic Techniques, Naval Reactors, Division of Reactor Development, United States Atomic Energy Commission (1964).

8. N. R. CANDELORE and R. C. GAST, "RECAP-3—A Monte Carlo Program for Estimating Epithermal Capture Rates in Rectangular or 60° Parallelogram Geometry," *WAPD-TM*-437 (March, 1964).

9. G. J. HABETLER, R. A. PFEIFFER, and L. E. TUECKE, "DRAM—A Monte Carlo Program for the Calculation of Capture Probabilities," *KAPL*-3013 (1964).

Here $\hat{f}$ is an upper bound† on $f(E, \boldsymbol{\omega})$, over all E and $\boldsymbol{\omega}$, and over all points, P, on τ. Now one generates a random number, ρ. If $R(E_0, \boldsymbol{\omega}) \leq \rho$, the starter is rejected and a new P and $\boldsymbol{\omega}$ are chosen. Otherwise the starter is accepted and given the weight

$$
W_1 = \int_\tau d\mathbf{r} \int_{\boldsymbol{\omega}\cdot\mathbf{n}>0} d\boldsymbol{\omega}' \int_{E_L}^{E_U} Q_p^+(E', \boldsymbol{\omega})\, dE'
$$
$$
= V_f \int_{E_L}^{E_U} \frac{\Sigma_{Tf}(E')}{E'} P_0(E')\, dE'.
$$

The remainder of the history is generated conventionally, and the absorption rate in the fuel, due to all starters, is estimated just as in any other Monte Carlo.

Still reasoning as in Sections 4.6 and 4.7, we are also led to a Monte Carlo procedure for the estimation of I_2. In the computation of I_2 we simulate the surface source

$$
Q_p^-(\boldsymbol{\omega}) = (\mu/4\pi)\frac{\Sigma_{Sf}(E)}{E}[1 - e^{-\Sigma_{Tf}(E)l}]/\Sigma_{Tf}(E), \qquad \mu = -\boldsymbol{\omega}\cdot\mathbf{n} > 0.
$$

Starting energies are drawn from the distribution

$$
\pi_2(E) \equiv \int_\tau d\mathbf{r} \int_{\boldsymbol{\omega}\cdot\mathbf{n}<0} Q_p^-(E, \boldsymbol{\omega}) \Big/ \int_\tau d\mathbf{r} \int_{\boldsymbol{\omega}\cdot\mathbf{n}<0} d\boldsymbol{\omega} \int_{E_L}^{E_U} Q_p^-(E', \boldsymbol{\omega})\, dE'
$$
$$
= \frac{\Sigma_{Sf}(E)}{E} P_0(E) \Big/ \int_{E_L}^{E_U} \frac{\Sigma_{Sf}(E')}{E'} P_0(E')\, dE'. \tag{6.2.11}
$$

Starting points, P, are again uniform over τ, but now $\boldsymbol{\omega}$'s are taken from an isotropic *incoming* distribution. Rejection proceeds just as before, but survivors are given the weight

$$
W_2 \equiv V_f \int_{E_L}^{E_U} \frac{\Sigma_{Sf}(E')}{E'} P_0(E')\, dE'.
$$

As in Section 4.6, all survivors are forced to scatter in the fuel. From now on, the histories proceed conventionally and absorption rates in the fuel are computed by any convenient technique.

The procedure outlined above is perfectly feasible, and often quite useful, but it can be refined in many ways. We shall describe various refinements in later Sections: one, however, is rather simple and we wish to discuss it here. In I_1 we have included contributions due to starters, drawn from $Q_p^+(\boldsymbol{\omega})$, which

---

† Note that for slabs $f(E_0, \boldsymbol{\omega}) \leq 2$ while for cylinders $f(E_0, \boldsymbol{\omega}) < 2.13$, so that we may choose $\hat{f} = 2$ in one case, and $\hat{f} = 2.13$ in the other. (See p. 148.)

are reabsorbed in the fuel on their first flights. Let the first-flight reabsorption rate be I_1^1 and the subsequent reabsorption rate be I_1'. Then

$$I_1 = I_1^1 + I_1'$$

$$A_f = \left[\int_{E_L}^{E_U} \frac{\Sigma_{af}(E')}{E'} P_0(E') \, dE' - I_1^1 \right] - I_1' + I_2.$$

However, it is easy to show that

$$\int_{E_L}^{E_U} \frac{\Sigma_{af}(E')}{E'} P_0(E') \, dE' - I_1^1 = \int_{E_L}^{E_U} \frac{\Sigma_{af}(E')}{E'} P^*(E') \, dE',$$

so that

$$A_f = \int_{E_L}^{E_U} \frac{\Sigma_{af}(E')}{E'} P^*(E') \, dE' - I_1' + I_2.$$

Here $P^*(E)$ is the probability that a starter of energy E, drawn from a uniform distribution in the fuel, will make its first collision elsewhere. To compute I_1', we take starting energies from the distribution

$$\pi_1'(E) = \frac{\Sigma_{Tf}(E)}{E} \left\{ P^*(E) + \frac{\Sigma_{Sf}(E)}{\Sigma_{Tf}(E)} [P_0(E) - P^*(E)] \right\} \Big/ C,$$

$$C \equiv \int_{E_L}^{E_U} \frac{\Sigma_{Tf}(E')}{E'} \left\{ \frac{\Sigma_{af}(E')}{\Sigma_{Tf}(E')} P^*(E') + \frac{\Sigma_{Sf}(E')}{\Sigma_{Tf}(E')} P_0(E') \right\} dE'.$$

Starting points and angles are chosen as before, but if a starter is absorbed on its first collision, it is discarded; it is not counted as a starter. One must choose a new P and a new $\boldsymbol{\omega}$ without changing the starting energy. All starters carry the weight $W_1' = CV_f$.

In the method we described earlier first-flight reabsorptions were treated by Monte Carlo. Now we are computing the first-flight reabsorption rate analytically and, as a consequence, we can expect some decrease in the variance of A_f. However, we must pay a price for this decrease in variance, since we must now compute $P^*(E)$. This quantity plays an important role in the theory of resonance absorption. It seems worth while, therefore, to consider carefully how P^* may be evaluated. The computation of P^* will be discussed in Section 6.3.

At this point we have arrived at a computational method whose properties are quite different from those of conventional Monte Carlo. In a conventional Monte Carlo calculation the relative variance in A_f becomes infinite as the resonance escape probability, P, goes to one. If we use superposition, the relative variance in A_f remains finite as P goes to one. The efficiency of the superposition method will be discussed in some detail later, in Section 6.4. We shall see there that the superposition method is usually most efficient when conventional Monte Carlo is least efficient.

6.3 COMPUTATION OF $P*$

We shall assume throughout this chapter that all fuel rods are identical, and that cross sections in the cladding and moderator are energy independent. It follows that $P*(E)$ is a well-defined function of the fuel cross section:

$$P*(E) = P*[\bar{l}\Sigma_{Tf}(E)] = P*(\bar{l}\Sigma_{Tf}).$$

Therefore, in using superposition, we need not tabulate $P*$ explicitly as a function of energy. Instead we may tabulate and store $P*$ as a function of the optical thickness of the fuel. Actually, we prefer to store $1/P*$, since the inverse of $P*$ is somewhat better behaved at the origin than $P*$. The reader will recall a similar convention in Section 4.4, where we discussed the tabulation of P_0.

It is our practice to tabulate $1/P*$ at 46 optical thicknesses, covering the range $0 \leq \bar{l}\Sigma_{Tf} \leq 12$. Normally we use the same optical thickness in tables of $1/P*$ as we have used to tabulate $1/P_0$.† We have always found that linear interpolation in such $1/P*$ tables is perfectly adequate for our purposes. Outside the range of the table

$$P*(\bar{l}\Sigma_{Tf}) \doteq 12\,P*(12)/\bar{l}\Sigma_{Tf}.$$

Of course, $P*(0) = 1$.

It seems, then, that we must evaluate $P*$, somehow, at 46 different optical thicknesses. This does not mean, however, that we need to compute 46 $P*$'s by Monte Carlo. Fortunately, $P*$ has properties which we can use to good advantage. One can show that

$$P*(\bar{l}\Sigma_{Tf}) \doteq P_0(\bar{l}\Sigma_{Tf})G_m/\{1 - (1 - G_m)[1 - G_f(\bar{l}\Sigma_{Tf})]\} \quad (6.3.1)$$

(see Ref. 1). Here G_m is the sticking probability of the cladding and moderator for an isotropic incident flux. In other words, G_m is the probability that an average neutron, impinging on the cladding and drawn from an isotropic flux, will collide before re-entering the fuel. Similarly, $G_f(\bar{l}\Sigma_{Tf})$ is the sticking probability of the *fuel* for an isotropic incident flux:

$$G_f(\bar{l}\Sigma_{Tf}) = \bar{l}\Sigma_{Tf}P_0(\bar{l}\Sigma_{Tf}).$$

We now define a quantity, γ_m, which exactly satisfies the following equation:

$$P*(\bar{l}\Sigma_{Tf}) = P_0(\bar{l}\Sigma_{Tf})\gamma_m(\bar{l}\Sigma_{Tf})/\{1 - [1 - \gamma_m(\bar{l}\Sigma_{Tf})][1 - G_f(\bar{l}\Sigma_{Tf})]\}.$$

$$(6.3.2)$$

Comparing Eqs. (6.3.1) and (6.3.2), we see that $\gamma_m(\bar{l}\Sigma_{Tf}) \doteq G_m$. Therefore, one may suspect that γ_m is not very sensitive to changes in the total cross

† See p. 155.

section, and we have found this to be true. Assuming that γ_m is, in fact, a slowly varying function of its argument, we proceed as follows. We compute $P^*(\bar{l}\Sigma_{Tfi})$ (by Monte Carlo) at the points $\bar{l}\Sigma_{Tfi} = 0, 1.2, 2.4, 3.6, \ldots, 12.0$. Thus, we construct a coarse table of P^*'s. Given $P^*(\bar{l}\Sigma_{Tfi})$ for any $\bar{l}\Sigma_{Tfi} \neq 0$, we can solve Eq. (6.3.2) for $\gamma_m(\bar{l}\Sigma_{Tfi})$. At $\bar{l}\Sigma_{Tfi} = 0$, Eq. (6.3.2) is indeterminate. We shall assume that $\gamma_m(\bar{l}\Sigma_{Tf}) = \gamma_m(1.2)$ for $0 \leq \bar{l}\Sigma_{Tf} \leq 1.2$, and that γ_m is linear in the intervals between all other mesh points. Under these assumptions we can compute $\gamma_m(\bar{l}\Sigma_{TfJ})$ at all points $\bar{l}\Sigma_{TfJ} = 0, 0.1, 0.2, \ldots,$ 2.0, 2.4, 2.8, $\ldots$, 12.0. Again using Eq. (6.3.2), we compute $P^*(\bar{l}\Sigma_{TfJ})$ and $1/P^*(\bar{l}\Sigma_{TfJ})$ at the same points. We now have our 46-point table, a table based on 10 P^*'s computed, as we have said, by Monte Carlo. But we have not yet discussed the Monte Carlo calculation itself. Now it is natural to ask: "How can we compute P^* efficiently via Monte Carlo?" One thing seems clear. We must compute all P^*'s simultaneously. This we do as follows.

We treat the cladding and moderator as pure absorbers without altering their total cross sections. Starters are distributed uniformly over the surface of the fuel. For each starter an ω is chosen from an isotropic distribution. Each starter is regarded as a composite of 10 particles and carries 10 weights, W_i. Initially

$$W_i^{(0)} = 2\mu[1 - e^{-\Sigma_{Tfi}l}]/\bar{l}\Sigma_{Tfi}P_0(\bar{l}\Sigma_{Tfi}),$$

where $\bar{l}\Sigma_{Tfi} = 1.2, 2.4, \ldots, 12.0$ for $i = 1, 2, \ldots, 10$. For each starter we pick an optical path length from an exponential distribution. The starter's path is then constructed as though the fuel rod were a vacuum. Suppose that the path traverses the fuel m times: let l_i be the chord length laid off in the fuel on the i-th crossing, and $L = \sum_{i=1}^{m} l_i$ be the total chord length in the fuel. Then the final weight of the i-th particle is

$$W_i^{(m)} = W_i^{(0)}e^{-\Sigma_{Tfi}L},$$

and the weight which it leaves in the fuel is

$$\Delta W_i = W_i^{(0)}[1 - e^{-\Sigma_{Tfi}L}].$$

It is easy to show that

$$P^*(\bar{l}\Sigma_{Tfi}) = P_0(\bar{l}\Sigma_{Tfi})E[1 - \Delta W_i].$$

In other words, the quantity $P_0(\bar{l}\Sigma_{Tfi})[1 - \Delta W_i]$ is an unbiased estimator of P^*, and we compute P^* by averaging this estimator over all starters. Again the reader will note that we must evaluate, repeatedly, the function $1 - e^{-x}$ and again we point out that this should be done via g-function tables (see Section 4.3).

6.4 ANALYSIS OF MODEL PROBLEM

In a realistic resonance escape problem it is very difficult to predict the variance in the computed escape probability. It is possible, however, to calculate this variance in a very simple "model problem". We concede that the superposition method is never as efficient in practice as it is in the model problem, and later we shall see why this is true. Still, analysis of the model problem shows us how the method works in ideal circumstances.

Suppose that there is no cladding, and that regions I and III are each so small that the cell may be considered homogeneous. We shall assume that there is no scattering in region I, while region III is filled with hydrogen. Since the cell is practically homogeneous,

$$P^*(\Sigma_{Tf}\bar{\bar{l}}) \doteq V_m\Sigma_{Tm}/[V_f\Sigma_{af} + V_m\Sigma_{Tm}],$$

where V_m and Σ_{Tm} are, respectively, the volume and total cross section of the moderator. We see that, approximately,

$$I_0 = V_m\Sigma_{Tm}\int_{E_L}^{E_U} \frac{\Sigma_{af}(E')V_f}{[\Sigma_{af}(E')V_f + \Sigma_{Tm}V_m]} \frac{dE'}{E'}.$$

Since there is no scattering in the fuel, $I_2 = 0$ and

$$A_f = V_m\Sigma_{Tm}\int_{E_L}^{E_U} \frac{\Sigma_{af}(E')V_f}{[\Sigma_{af}(E')V_f + \Sigma_{Tm}V_m]} \frac{dE'}{E'} - I_1'.$$

Now, we have assumed that the moderator is hydrogen, and that $\phi(E) = 1/E$ above the resonances. Under such circumstances one can show that the total number of neutrons, Q, entering the resonance range is equal to $V_m\Sigma_{Tm}$. Since P, the resonance escape probability, is equal to $(1 - A_f/Q)$,

$$1 - P = \int_{E_L}^{E_U} \frac{\Sigma_{af}(E')V_f}{[\Sigma_{af}(E')V_f + \Sigma_{Tm}V_m]} \frac{dE'}{E'} - I_1'/V_m\Sigma_{Tm}. \qquad (6.4.1)$$

But, for pure hydrogen moderation,†

$$P = \exp\left\{-\int_{E_L}^{E_U} \frac{\Sigma_{af}(E')V_f}{[\Sigma_{af}(E')V_f + \Sigma_{Tm}V_m]} \frac{dE'}{E'}\right\} \qquad (6.4.2)$$

(Ref. 2). Consequently,

$$I_1' = -V_m\Sigma_{Tm}[1 - P + \ln(P)] \qquad (6.4.3)$$

and we see that, in the model problem, I_1' is very simply related to P. It follows, as we now show, that the variance in the escape probability is also determined by P.

---

† Here we are neglecting absorption in hydrogen.

In the computation of I_1' all starters carry the weight

$$W_1' = V_f \int_{E_L}^{E_U} \frac{\Sigma_{af}(E')}{E'} P^*(E')\, dE' = -V_m \Sigma_{Tm} \ln (P).$$

I_1' itself is the average weight reabsorbed in the fuel. If p_r is the probability that an average starter will be reabsorbed, then

$$I_1' = W_1' p_r = -V_m \Sigma_{Tm} p_r \ln (P),$$

$$p_r = -I_1'/V_m \Sigma_{Tm} \ln (P). \tag{6.4.4}$$

Combining Eqs. (6.4.3) and (6.4.4), we see that

$$p_r = [1 - P + \ln (P)]/\ln (P) = 1 + (1 - P)/\ln (P). \tag{6.4.5}$$

The reader may observe that the Monte Carlo process we have devised to compute I_1' could be regarded as a method for computing p_r. Suppose that we choose to estimate p_r by simply counting reabsorbed starters. Then the variance in p_r will be given by the expression

$$\sigma^2(p_r) = p_r(1 - p_r) = -(1 - P)[1 + (1 - P)/\ln (P)]/\ln (P).$$

Further, from (6.4.1),

$$\sigma^2(1 - P, \text{superposition}) = \sigma^2[I_1'/\Sigma_{Tm} V_m]$$
$$= \sigma^2[p_r \ln (P)],$$

or

$$\sigma^2(1 - P, \text{superposition}) = -(1 - P)[\ln (P) + (1 - P)]. \tag{6.4.6}$$

Equation (6.4.6) is the desired relation between variance and escape probability. As for the relative variance,

$$\sigma_{r,s}^2 \equiv \frac{\sigma^2(1 - P, \text{superposition})}{(1 - P)^2} = -[1 + \ln (P)/(1 - P)]. \tag{6.4.7}$$

This expression for the relative variance exhibits quite clearly a most important feature of the superposition method. It will be seen that

$$\sigma_{r,s}^2 \begin{cases} = 0 & \text{when} \quad P = 1, \\ = \infty & \text{when} \quad P = 0. \end{cases}$$

In other words, the superposition method is most useful when P is large. On the other hand, in a conventional calculation†

$$\sigma_{r,c}^2 \equiv \frac{\sigma^2(1 - P, \text{conventional})}{(1 - P)^2} = \frac{P}{(1 - P)},$$

---

† Again we are assuming a binomial estimator.

so that

$$\sigma_{r,c}^2 \begin{cases} = 0 & \text{when} \quad P = 0, \\ = \infty & \text{when} \quad P = 1. \end{cases}$$

Thus, conventional Monte Carlo is most useful when P is small.

To compare the two methods more closely, we need some rough measure of the efficiency of each. Let $N_s(\sigma)$ be the number of histories required to achieve a given variance via superposition. Correspondingly, take $N_c(\sigma)$ to be the number of histories required to achieve the same variance by conventional methods. Then, in some sense, the ratio $\rho \equiv N_c/N_s$ is a ratio of efficiencies. In Table 6.1 we list ρ's for various P's. Table 6.1 shows that $\rho > 1$ so long as $P > e^{-1}$. In the range $P > 0.7$ (an important range for the reactor physicist) ρ is substantially greater than 10.

TABLE 6.1
THE RATIO ρ AS A FUNCTION OF RESONANCE ESCAPE PROBABILITY

P	e^{-1}	0.4	0.5	0.6	0.7	0.75	0.8	1
ρ	1	1.26	2.69	5.40	12.3	19.7	34.7	∞

It is clear that, for P sufficiently close to one, the superposition method will be overwhelmingly more efficient than conventional Monte Carlo. There is little more we can say unless we know something about the running time per history in computations of both types. Obviously, the superposition method involves a good deal of work at the beginning of each history. However, there is one advantageous feature of superposition which we have not yet mentioned. The density functions, $\pi_1'(E)$ and $\pi_2(E)$, from which we draw starting energies each contain $1/E$ as a factor. For this reason starters, in superposition calculations, tend to come predominantly from low energies. In conventional Monte Carlo the energy distribution of starters is proportional to the slowing-in density from above some energy E_U. The slowing-in density produced by hydrogen scattering is uniform in energy. Scattering by heavier elements gives rise to starters with energy distributions concentrated near E_U. Therefore, we usually find that the number of collisions per history is smaller in superposition calculations than in conventional Monte Carlo calculations. All things considered, we see no *a priori* reason why superposition calculations should run longer, per history, than conventional calculations, nor have we found this to be true in practice.

Still there is, at this point, a serious weakness in our analysis. In computing variances we have so far neglected all scattering in the fuel. It is easy to see that scattering in the fuel can cause us a good deal of trouble. Scattering will be troublesome if I_1' and I_2 are much larger than A_f. We show now that this will often be true.

The reader will recall that W_2, the total weight of starters in the computation of I_2, is equal to

$$V_f \int_{E_L}^{E_U} \frac{\Sigma_{Sf}(E')}{E'} P_0(E') \, dE'.$$

Further,

$$W_1' = V_f \int_{E_L}^{E_U} \left\{ \frac{\Sigma_{Tf}(E')}{E'} P^*(E') + \frac{\Sigma_{Sf}(E')}{E'} [P_0(E') - P^*(E')] \right\} dE'.$$

However, in most resonance escape calculations, the potential scattering cross section is taken to be energy independent. Since

$$\lim_{E_U \to \infty} \Sigma_{Tf}(E) \geq \lim_{E_U \to \infty} \Sigma_{Sf}(E) > 0,$$

it follows that

$$\lim_{E_U \to \infty} W_1' = \lim_{E_U \to \infty} W_2 = \infty.$$

Let us, for convenience, designate the Monte Carlo computation of I_1' as Monte Carlo 1, and the computation of I_2 as Monte Carlo 2. Let p_{r1} be the probability that an average starter in Monte Carlo 1 will ultimately be captured. Define the corresponding probability p_{r2}. It is easy to see that

$$\lim_{E_U \to \infty} p_{r1} > 0, \qquad \lim_{E_U \to \infty} p_{r2} > 0.$$

But

$$I_1' = p_{r1} W_1', \qquad I_2 = p_{r2} W_2,$$

and, in consequence,

$$\lim_{E_U \to \infty} I_1' = \infty = \lim_{E_U \to \infty} I_2.$$

On the other hand, the absorption rate in the fuel must certainly be finite, i.e.

$$\lim_{E_U \to \infty} A_f < \infty.$$

Thus, if the upper energy cut-point is high, we will find that

$$I_1' \gg A_f, \qquad I_2 \gg A_f.$$

Whenever I_1' and I_2 are much greater than A_f, the computation of A_f will be difficult, for

$$\sigma^2(A_f) = \sigma^2(I_2 - I_1').$$

Since Monte Carlo 1 and 2 are independent,

$$\sigma^2(I_2 - I_1') = \sigma^2(I_1') + \sigma^2(I_2)$$

and

$$
\frac{\sigma^2(A_f)}{A_f^2} = \frac{\sigma^2(I_1')}{A_f^2} + \frac{\sigma^2(I_2)}{A_f^2}
$$

$$
\gg \frac{\sigma^2(I_1')}{(I_1')^2} + \frac{\sigma^2(I_2)}{I_2^2}.
$$

Thus, the relative variance in A_f will be much greater than the relative variance in I_1' or I_2.

We see that the superposition method, as formulated above, is inefficient in computations covering large energy ranges. Unfortunately, it is not always efficient, either, in treating small energy ranges. In any range where

$$
\int_{E_L}^{E_U} \frac{\Sigma_{af}(E')}{E'} \, dE' < \int_{E_L}^{E_U} \frac{\Sigma_{Sf}(E')}{E'} \, dE'
$$

we are likely to find that I_1' or I_2, or both, are larger than A_f. The above inequality may hold even if the range of integration is as narrow as a single resonance: it will hold if the resonance scattering is sufficiently strong. How, then, are we to use the superposition method?

Despite its drawbacks, the method has some value, even in its present form. One can sometimes use it, for example, to treat uranium fuel highly enriched with U^{235}. In such highly enriched fuel most absorption usually takes place below 100 eV. At energies below 100 eV the scattering rate in U^{235} is small compared to the rate of absorption. Furthermore, we can with superposition compute $(1 - P)$ for a single resonance in an isolated fuel rod. In practice it is very nearly impossible to do this with conventional Monte Carlo. Clearly, if the moderator is very wide and the resonance fairly narrow, P will be close to one. In conventional Monte Carlo only a small fraction of all sample neutrons will pass through the fuel or contribute in any way to our estimate of the absorption rate, yet, if we use superposition, the width of the moderator and the narrowness of the resonance cause no difficulties. In computing absorption rates for individual resonances and isolated rods superposition is vastly superior to conventional methods. This is true even when resonance scattering is very strong.

We have developed superposition here as an alternative to the adjoint method. It is our opinion that the adjoint method is not well suited to resonance escape calculations, since it involves the use of varying weights.† However, by invoking the superposition principle instead of the reciprocity relation, we have not completely avoided all weight variations: a sample neutron absorbed in Monte Carlo 1 makes a negative contribution to A_f,

---

† Again we note that varying weights, judiciously constructed, may reduce the variance in Monte Carlo calculations but that weight variations *generally* tend to increase the variance.

while in Monte Carlo 2 each absorption makes a positive contribution to A_f. One cannot eliminate such weight variations, and in a sense this is the basic weakness in the superposition method. We see the basic *advantage* of superposition if we contrast the behavior of weights in the superposition and adjoint methods. In superposition calculations sample neutrons are assigned weights at birth and the starting weights are unaltered throughout the neutron's history. In adjoint resonance escape calculations the weight fluctuates from collision to collision. The neutron weights are not bounded, either from above or below. We must anticipate that the occasional arrival of very high weight neutrons in the fuel will raise havoc with all our estimators.

Clearly, any scattering in the fuel will reduce the efficiency of the superposition method in its present form. Throughout this chapter scattering in the fuel has been our most serious problem: it will remain our most serious problem but we shall try, in various ways, to improve the treatment of scattering. To do this we must make some basic changes in the superposition method.

6.5 SECOND METHOD†

Reasoning as in Section 4.5, we can interpret the superposition method as a perturbation method. If we analyze the method in its present form, we see that the unperturbed flux is equal to $1/E$ and that all collisions in the fuel are regarded as perturbations. In the "unperturbed state" the fuel is a vacuum, but it is not necessary to treat *all* collisions in the fuel as perturbations. It is possible to define an unperturbed state somewhat differently. We take the view, now, that the "unperturbed" fuel is a pure scatterer with no resonances. In the unperturbed state all resonance widths are set to zero but the potential scattering cross sections in the fuel are unaltered. Again, by definition, potential scattering cross sections are energy independent, and again we suppose that all scattering is isotropic in the center of mass. Under such conditions the unperturbed asymptotic flux (i.e. the flux at energies far below the energies of any sources) is still equal to $1/E$. Note that, in view of our new definition of the unperturbed state, we can improve on assumptions about the cross sections above E_U. It is an essential feature of the superposition method that the fuel, above E_U, is taken to be in its unperturbed state. Therefore, we will no longer assume that all fuel cross sections vanish at energies higher than E_U. Instead we assume only that there are no resonances, and that there is no absorption, at such energies. Each element in the fuel is allowed to have an energy independent potential cross section both above and below E_U. Having changed our definition of the unperturbed

---

† What we call the "second method" here, is a combination of the first and second methods of Chapter 4.

state, we set out now to make corresponding changes in the superposition method. To do this we shall have to revise problem β. In our new problem β we put a volume source,

$$Q_f(E) = \frac{1}{E}[\Sigma_{af}(E) + \Sigma_{Srf}(E)] \equiv \frac{1}{E}\Sigma_{rf}(E), \qquad (6.5.1)$$

in the fuel, while, again, the source density outside the fuel is zero. At this point we have introduced a cross section, $\Sigma_{Srf}(E)$, which will be used a good deal in our work. For the i-th isotope in region I,

$$\Sigma^i_{Srf}(E) \equiv \Sigma^i_{Sf}(E) - \Sigma^i_{Pf}.$$

Here Σ^i_{Sf} and Σ^i_{Pf} are, respectively, the total and potential scattering cross sections of the isotope. We assume that the potential scattering cross section is energy independent. Of course, $\Sigma_{Srf}(E)$ is the sum of all the resonance scattering cross sections in the region, $\Sigma_{Srf}(E) = \sum_i \Sigma^i_{Srf}(E)$.[†]
  Again we let $F_T = F_\alpha + F_\beta$. Define an operator $\mathscr{L}$ such that

$$\mathscr{L}F(E, \mathbf{r}, \boldsymbol{\omega}) = \boldsymbol{\omega} \cdot \nabla F(E, \mathbf{r}, \boldsymbol{\omega}) + \Sigma_T(E, \mathbf{r})F(E, \mathbf{r}, \boldsymbol{\omega}). \qquad (6.5.2)$$

Let $F_{TU}(E, \mathbf{r}, \boldsymbol{\omega})$ be a flux which satisfies the transport equation

$$\mathscr{L}F_{TU}(E, \mathbf{r}, \boldsymbol{\omega}) = \int d\boldsymbol{\omega}' \int_{E_L}^{E_U} dE' \Sigma_{Pf}(\boldsymbol{\omega} \cdot \boldsymbol{\omega}', E' \to E, \mathbf{r})F_{TU}(E', \mathbf{r}, \boldsymbol{\omega}')$$
$$+ (1/4\pi)Q_S(E, \mathbf{r}) + (1/4\pi)Q_f(E), \qquad \mathbf{r} \in R_\mathrm{I},$$

$$\mathscr{L}F_{TU}(E, \mathbf{r}, \boldsymbol{\omega}) = \int d\boldsymbol{\omega}' \int_{E_L}^{E_U} dE' \Sigma_S(\boldsymbol{\omega} \cdot \boldsymbol{\omega}', E' \to E, \mathbf{r})F_{TU}(E', \mathbf{r}, \boldsymbol{\omega}')$$
$$+ (1/4\pi)Q_S(E, \mathbf{r}), \qquad \mathbf{r} \notin R_\mathrm{I}. \quad (6.5.3)$$

In Eq. (6.5.3) the quantity $Q_S(E, \mathbf{r})$ is the slowing-in density from energies above E_U. More precisely, $Q_S(E, \mathbf{r})\,dE$ is the rate at which neutrons slow down, per unit volume, from above E_U into an interval dE in the neighborhood of E.
  All scattering cross sections which appear in (6.5.3) are constant. There is no absorption except in region I, where the source is equal to $1/4\pi E$ times the absorption cross section.[‡] If $F(E, \mathbf{r}, \boldsymbol{\omega}) = 1/4\pi E$ for $E > E_U$, it follows that

$$F_{TU}(E, \mathbf{r}, \boldsymbol{\omega}) = 1/4\pi E \qquad (6.5.4)$$

in the range $E_L \leq E \leq E_U$.

---

† Note that the interference scattering cross section may be negative and that, therefore, Σ_{Srf} may be negative at some energies.
‡ In Eq. (6.5.3) we are, in effect, treating resonance scattering as though it were absorption.

Now define $F_{TS}(E, \mathbf{r}, \boldsymbol{\omega})$ so that

$$\mathscr{L} F_{TS}(E, \mathbf{r}, \boldsymbol{\omega}) = \int d\boldsymbol{\omega}' \int_{E_L}^{E_U} dE' \Sigma_{Sf}(\boldsymbol{\omega} \cdot \boldsymbol{\omega}', E' \to E, \mathbf{r}) F_{TS}(E', \mathbf{r}, \boldsymbol{\omega}') \\ + Q_{TS}(E, \mathbf{r}, \boldsymbol{\omega}), \quad (6.5.5)$$

$$Q_{TS}(E, \mathbf{r}, \boldsymbol{\omega}) = \begin{cases} \int \int d\boldsymbol{\omega}' \int_{E_L}^{E_U} dE' \Sigma_{Srf}(\boldsymbol{\omega} \cdot \boldsymbol{\omega}', E' \to E, \mathbf{r}) F_{TU}(E', \mathbf{r}, \boldsymbol{\omega}'), \\ \hspace{7cm} \mathbf{r} \in R_I \\ 0, \quad \mathbf{r} \notin R_I. \end{cases}$$

Adding Eqs. (6.5.3) and (6.5.5), we see that $F_T = F_{TU} + F_{TS}$, and, thus,

$$F_\alpha = F_{TU} + F_{TS} - F_\beta = (1/4\pi E) + F_{TS} - F_\beta. \quad (6.5.6)$$

We proceed now much as we did in Section 4.7, but here we must modify slightly our definitions of the components of F_β. It will be convenient to refer to Σ_P and Σ_{Sr} as cross sections for different types of events, events which we designate as "potential scatterings" and "resonance scatterings." Let

$$F_\beta = F_\beta \text{ (unscattered before exit, unreturned)}$$
$$+ F_\beta \text{ (unscattered before exit, returned)}$$
$$+ F_\beta \text{ (resonance scattered before exit)}$$
$$+ F_\beta \text{ (potential scattered before exit)}.$$

More briefly,

$$F_\beta = F_{\beta uu} + F_{\beta ur} + F_{\beta Sr} + F_{\beta SP}.$$

In this abbreviated notation

$$F_\alpha = (1/4\pi E) - F_{\beta uu} - F_{\beta ur} - F_{\beta SP} + F_{TS} - F_{\beta Sr}.$$

As in Section 6.2,

$$A_f \equiv \int_{E_L}^{E_U} dE' \int_I d\mathbf{r} \Sigma_{af}(E') \phi_\alpha(E', \mathbf{r}).$$

If we write

$$I_0 \equiv \int_{E_L}^{E_U} dE' \int_I d\mathbf{r} \Sigma_{af}(E')[(1/E') - \phi_{\beta uu}(E', \mathbf{r})], \quad (6.5.7)$$

$$I_1 \equiv \int_{E_L}^{E_U} dE' \int_I d\mathbf{r} \Sigma_{af}(E') \phi_{\beta ur}(E', \mathbf{r}), \quad (6.5.8)$$

$$I_2 \equiv \int_{E_L}^{E_U} dE' \int_I d\mathbf{r} \Sigma_{af}(E') \phi_{\beta SP}(E', \mathbf{r}), \quad (6.5.9)$$

$$J \equiv \int_{E_L}^{E_U} dE' \int_I d\mathbf{r} \Sigma_{af}(E')[\phi_{TS}(E', \mathbf{r}) - \phi_{\beta Sr}], \quad (6.5.10)$$

then
$$A_f = I_0 - I_1 - I_2 + J. \qquad (6.5.11)$$

Now, through simple algebraic manipulation, we find that

$$(1/E) - \phi_{\beta uu}(E, \mathbf{r}) = [Q_f(E) - \Sigma_{Tf}(E)\phi_{\beta uu}(E, \mathbf{r})]/\Sigma_{rf}(E)$$
$$+ \Sigma_{Tf}(E)\Sigma_{Pf}\phi_{\beta uu}(\mathbf{r})/\Sigma_{Tf}(E)\Sigma_{rf}(E).$$

Consequently,

$$\int_I d\mathbf{r}\Sigma_{af}(E)[(1/E) - \phi_{\beta uu}(E, \mathbf{r})]$$
$$= V_I\Sigma_{af}(E)\left\{P_0(E) + \frac{\Sigma_{Pf}}{\Sigma_{Tf}(E)}[1 - P_0(E)]\right\}\Big/E,$$
$$I_0 = V_I\int_{E_L}^{E_U}dE'\Sigma_{af}(E')\left\{P_0(E') + \frac{\Sigma_{Pf}}{\Sigma_{Tf}(E')}[1 - P_0(E')]\right\}\Big/E'.$$

It will be seen that I_0 can be evaluated by deterministic quadrature methods.

To compute $-I_1$ is not difficult. We can show, by quite familiar arguments, that I_1 is the absorption rate produced by a source

$$Q_{1P}(E, \boldsymbol{\omega}) = \mu\Sigma_{rf}(E)[1 - e^{-l\Sigma_{Tf}(E)}]/4\pi E\Sigma_{Tf}(E), \qquad \mu \equiv \boldsymbol{\omega} \cdot \mathbf{n} > 0$$
$$= 0, \qquad \mu < 0 \qquad (6.5.12)$$

distributed over τ. This source may be negative at some energies,† but otherwise it is essentially the same as others we have encountered earlier. Using techniques already discussed, we may sample this source, follow the starters, and estimate the absorption rate in the fuel.

The computation of $-I_2$ is also straightforward. In problem β the number of source neutrons produced, per second, at E is equal to $V_I Q_f(E)$. Therefore, if $Q_2(E)$ is the rate at which unreturned neutrons undergo potential scattering, then

$$Q_2(E) = V_I |\Sigma_{rf}(E)| \Sigma_{Pf}[1 - P_0(E)]/E\Sigma_{Tf}(E).$$

To compute I_2 we

a) Select starting energies from the density function

$$\pi_2(E) = Q_2(E)\Big/\int_{E_L}^{E_U}Q_2(E')\,dE'.$$

b) Give each starter the weight $-W_2 = \int_{E_L}^{E_U} Q_2(E')\,dE'$ if $\Sigma_{rf}(E)$ is positive. Otherwise the weight is equal to $+W_2$.

---

† See footnote to p. 217.

c) Draw starting points from a distribution uniform over V_I, and starting $\boldsymbol{\omega}$'s from an isotropic distribution. If a starter leaves V_I on its first flight we reject it. Otherwise we regard its first collision as a potential scattering collision and generate the rest of its history conventionally.

It is clear that, if $[1 - P_0(E)]$ is small at some energy, then, at that energy, our rejection process will be very inefficient. However, this will usually happen at energies, far from resonance peaks, where $\pi_2(E)$ is likely to be small. Such energies are rarely drawn from the density function.†

We have now just one more integral to compute but, unfortunately, this integral is somewhat more troublesome than the others. Let us define $F_{TS} - F_{\beta Sr} = F_{\Delta S}$. It is easy to show, by arguments like those in Section 4.7, that

$$\mathscr{L} F_{\Delta S}(E, \mathbf{r}, \boldsymbol{\omega}) = \int_{E_L}^{E_U} dE' \int d\boldsymbol{\omega}' \Sigma_{Sf}(\boldsymbol{\omega} \cdot \boldsymbol{\omega}', E' \to E, \mathbf{r}) F_{\Delta S}(E', \mathbf{r}, \boldsymbol{\omega}')$$
$$+ Q_{\Delta S}(E, \mathbf{r}, \boldsymbol{\omega}), \quad (6.5.13)$$

$$Q_{\Delta S}(E, \mathbf{r}, \boldsymbol{\omega}) = \begin{cases} \int_{E_L}^{E_U} dE' \int d\boldsymbol{\omega}' \Sigma_{Srf}(\boldsymbol{\omega} \cdot \boldsymbol{\omega}', E' \to E, \mathbf{r})[F_{TU}(E', \mathbf{r}, \boldsymbol{\omega}') \\ \qquad\qquad - F_{\beta uu}(E', \mathbf{r}, \boldsymbol{\omega}')], \quad \mathbf{r} \in R_I, \\ 0, \quad \mathbf{r} \notin R_I. \end{cases} \quad (6.5.14)$$

Since $F_{TU}(E, \mathbf{r}, \boldsymbol{\omega}) = 1/4\pi E$, it is clear that

$$\mathscr{L} F_{TU}(E, \mathbf{r}, \boldsymbol{\omega}) = \Sigma_{Tf}(E)/4\pi E, \quad \mathbf{r} \in R_I,$$
$$F_{TU}(E, \mathbf{r}_p, \boldsymbol{\omega}) = 1/4\pi E. \quad (6.5.15)$$

Further, by definition,

$$\mathscr{L} F_{\beta uu}(E, \mathbf{r}, \boldsymbol{\omega}) = \Sigma_{rf}(E)/4\pi E, \quad \mathbf{r} \in R_I,$$
$$F_{\beta uu}(E, \mathbf{r}_p, \boldsymbol{\omega}) = 0, \quad \boldsymbol{\omega} \cdot \mathbf{n} < 0. \quad (6.5.16)$$

Subtracting (6.5.16) from (6.5.15):

$$\mathscr{L}[F_{TU}(E, \mathbf{r}, \boldsymbol{\omega}) - F_{\beta uu}(E, \mathbf{r}, \boldsymbol{\omega})] = \Sigma_{Pf}/4\pi E, \quad \mathbf{r} \in R_I,$$
$$\quad (6.5.17)$$
$$F_{TU}(E, \mathbf{r}_p, \boldsymbol{\omega}) - F_{\beta uu}(E, \mathbf{r}_p, \boldsymbol{\omega}) = 1/4\pi E, \quad \boldsymbol{\omega} \cdot \mathbf{n} < 0.$$

We see from Eq. (6.5.17) that the quantity $[F_{TU}(E, \mathbf{r}, \boldsymbol{\omega}) - F_{\beta uu}(E, \mathbf{r}, \boldsymbol{\omega})]$ satisfies a transport equation in I, and that this quantity can be regarded as a flux produced by the joint action of two sources. One is a uniform, isotropic, volume source of density $\Sigma_{Pf}/4\pi E$, while the other is an inward-directed surface source with strength equal to $\boldsymbol{\omega} \cdot \mathbf{n}/4\pi E$.

---

† It is for this reason that the inefficiency of the rejection process when $P_0 \doteq 1$ is not as serious in multienergy as in one-energy calculations. That the weakness of the rejection process is not very important has already been noted (p. 167).

At this point we arrive at the following conclusions. From Eq. (6.5.10) we know that J is the absorption rate induced in R_I by the flux $F_{\Delta S}$. It is clear from (6.5.13) and (6.5.14) that $F_{\Delta S}$ is produced by neutrons resonance-scattered, within R_I, in the flux $(F_{TU} - F_{\beta uu})$. Therefore, to compute J, we must

a) generate the flux $(F_{TU} - F_{\beta uu})$,
b) generate resonance scattering events in this flux,
c) track the scattered neutrons, and, finally,
d) estimate the rate at which these scattered neutrons are absorbed in the fuel.

To carry out steps (a) through (d) we may, for example, invoke two separate Monte Carlo procedures, described below.

First define

$$Q_3(E) = V_\mathrm{I}\Sigma_{Pf}\,|\Sigma_{Srf}(E)|\,[1 - P_0(E)]/E\Sigma_{Tf}(E),$$
$$\pi_3(E) = Q_3(E)\Big/ \int_{E_L}^{E_U} Q_3(E)\,dE. \tag{6.5.18}$$

We draw starting energies from $\pi_3(E)$. If $\Sigma_{Srf}(E) > 0$, we give the starter the weight $W_3 = \int_{E_L}^{E_U} Q_3(E)\,dE$. Otherwise the weight is equal to $-W_3$. Starting points, P, are uniform over R_I and starting $\boldsymbol{\omega}$'s are isotropic. If a starter leaves R_I on its first flight, it is discarded. Without changing E we take a new P and a new $\boldsymbol{\omega}$. Otherwise we follow the starter, force it to resonance-scatter, and estimate the absorption rate it produces in R_I. Let the absorption rate due to starters draw from $Q_3(E)$ be I_3.

Next we define an ingoing surface source

$$Q_{4P}^-(E, \boldsymbol{\omega}) = \mu|\Sigma_{Srf}(E)|\,[1 - e^{-l\Sigma_{Tf}(E)}\,]/4\pi E\Sigma_{Tf}(E),$$
$$\mu \equiv -\boldsymbol{\omega}\cdot\mathbf{n} > 0. \tag{6.5.19}$$

One can show that

$$Q_4(E) \equiv \int_\tau d\mathbf{r} \int_{\boldsymbol{\omega}\cdot\mathbf{n}<0} d\boldsymbol{\omega}\,|Q_{4P}^-(E, \boldsymbol{\omega})| = |\Sigma_{Srf}(E)|\,V_\mathrm{I}P_0(E)/E. \tag{6.5.20}$$

Define

$$\pi_4(E) = Q_4(E)\Big/ \int_{E_L}^{E_U} Q_4(E')\,dE'. \tag{6.5.21}$$

Draw starting energies from $\pi_4(E)$, and starting points from a distribution uniform over τ. Give each starter the weight

$$W_4 = \int_{E_L}^{E_U} Q_4(E')\,\mathrm{sgn}\,[\Sigma_{Srf}(E')]\,dE'.$$

Draw ω's from an isotropic distribution. Compute

$$f(E, \omega) = Q_{4P}^-(E, \omega)/(1/2\pi)(1/A_I)\int_\tau d\mathbf{r}\int_{\omega \cdot \mathbf{n} < 0} d\omega Q_{4P}^-(E, \omega)$$

$$= 2\mu[1 - e^{-l\Sigma_{Tf}(E)}]/\bar{l}\Sigma_{Tf}(E)P_0(E).$$

Form $R \equiv |f(E, \omega)|/\bar{f}$. Pick a random number, ρ, and reject the starter if $R < \rho$. Otherwise force a resonance scattering collision in R_I and follow the subsequent history conventionally. Let the absorption rate due to starters drawn from Q_4 be I_4. It will be seen that $A_f = I_0 - I_1 - I_2 + I_3 + I_4$.

We observe that we have included, in I_1, contributions due to starters which have been absorbed on their first collision. As in Section 6.2 we prefer, here also, to delete such contributions from I_1 and include them in I_0. Let us again write $I_1 \equiv I_1^1 + I_1'$. It is easy to show that

$$I_1^1 = V_I \int_{E_L}^{E_U} dE' \Sigma_{rf}(E')\Sigma_{af}(E')[P_0(E') - P^*(E')]/E'\Sigma_{Tf}(E').$$

Therefore,

$$I_0' \equiv I_0 - I_1^1 = V_I \int_{E_L}^{E_U} \frac{dE'}{E'} \Sigma_{af}(E')\left\{P^*(E') + \frac{\Sigma_{Pf}}{\Sigma_{Tf}(E')}[1 - P^*(E')]\right\}.$$
$$\tag{6.5.22}$$

The reader who is familiar with resonance escape theory will recognize that I_0' is the resonance absorption rate in the NR approximation (Ref. 2). Of course,

$$A_f = I_0' - I_1' - I_2 + I_3 + I_4. \tag{6.5.23}$$

Thus, in computing I_1', I_2, I_3, and I_4, we are simply correcting the NR approximation.

Unfortunately, it does not follow that the Monte Carlo corrections will all be small if the NR approximation is accurate. It remains true here, as in Section 6.2, that the Monte Carlo corrections are not all of one sign. The individual corrections may be large even if their sum is small. What, then, have we gained through our separate treatment of potential and resonance scattering? We argue that we have, in fact, gained a great deal. Earlier, in Section 6.4, we have analyzed the performance of our First Method when it is applied to a simple model problem. Our analysis showed that the method is very efficient, but the analysis becomes invalid if there is *any* scattering in the fuel. Any scattering in the fuel lowers the efficiency of the method.

Let us suppose, however, that there is no *resonance*-scattering in the fuel. Then, in applying the Second Method, we find that $I_3 = I_4 = 0$. Further, if region I is infinitely small, $I_2 = 0$. Consequently, all the reasoning of Section 6.4 is again valid, including the reasoning which led us to Table 6.1. Potential scattering does not impair the efficiency of the Second Method: only resonance-scattering remains troublesome.

6.6 STORAGE-SAVING TECHNIQUES

We have implied, in Section 6.5, that the various Monte Carlo corrections are to be computed separately. While this may be done, such an approach is somewhat wasteful of storage. If we compute I_1', I_2, I_3, and I_4 separately, we must store, separately, tables of the functions

$$F_1'(E) = \int_{E_L}^{E} \pi_1'(E')\,dE', \qquad\qquad F_2(E) = \int_{E_L}^{E} \pi_2(E')\,dE',$$

$$F_3(E) = \int_{E_L}^{E} \pi_3(E')\,dE', \quad \text{and} \quad F_4(E) = \int_{E_L}^{E} \pi_4(E')\,dE'.$$

However, it is possible to compute all four integrals simultaneously and to store only a single distribution function. We proceed as follows.

Define

$$Q_1(E) = |\Sigma_{rf}(E)|\,\{P_0(E) - \Sigma_{af}(E)[P_0(E) - P^*(E)]/\Sigma_{Tf}(E)\}/E,$$

$$Q(E) = Q_1(E) + Q_2(E) + Q_3(E) + Q_4(E),$$

$$\pi(E) = Q(E) \Big/ \int_{E_L}^{E_U} Q(E)\,dE',$$

and

$$W = \int_{E_L}^{E_U} Q(E')\,dE'.$$

Take a starting energy from the density $\pi(E)$. Divide all starters into four categories. Let $Q_i(E)/Q(E)$ be the probability that a starter, at energy E, will fall into the i-th category. Choose a category. Suppose that the first category has been chosen. Then, if $\Sigma_{rf}(E) > 0$, the starter is assigned a weight equal to $-W$. Otherwise the weight is equal to W. Starting angles and positions are selected as in the computation of I_1'. Suppose the second category has been chosen. Then, if $\Sigma_{Srf}(E) > 0$, the starting weight is $-W$. Otherwise it is equal to W. Starting angles and positions are determined as in the computation of I_2. In categories 3 and 4 the weights are equal to W if $\Sigma_{Srf}(E) > 0$, and equal to $-W$ otherwise. Starting parameters are chosen as in the computations of I_3 and I_4, respectively. Contributions of all sample histories to the region I absorption rate are combined, taking note of the various starters.

Of course, there are other techniques which may be used, in Monte Carlo, to minimize storage requirements, and there is one in particular which we have found rather helpful. This technique allows us to avoid storing large numbers of scattering cross sections. If an isotope has no resonances, its scattering cross section (according to assumptions we have made) is energy independent. Therefore, it is easy to store scattering cross sections for many such isotopes. However, if an isotope has resonances its scattering is a complicated function of energy. A table of resonance scattering cross sections is, generally, very large and unwieldy, and occupies large storage blocks.

Now, it is often true that, in a given problem configuration, all isotopes with resonances have approximately the same mass. They may, for example, all be isotopes of uranium or thorium. In such a situation we make no serious error if we assume that they have *precisely* the same mass. One may take, as a fictitious mass, some average of the masses of the various resonance isotopes. Having neglected mass differences, it is unnecessary to distinguish the scattering cross sections of the different isotopes, and we store only a single table of resonance-scattering cross sections.

6.7 UTILITY OF THE SUPERPOSITION METHOD

At this time we see little more that one can do to improve the superposition method. Therefore, it may be appropriate now to consider how this method can be used to best advantage, and to examine its capabilities and limitations.

Apparently it has two principal limitations. First of all, the super-position method is not very useful when P, the resonance escape probability, is small. In problems where $P \leq \frac{1}{2}$ other computational techniques are preferable. It was not to treat such problems that the method was devised.

Secondly, the superposition method is inefficient if the resonance scattering rate is high compared to the absorption rate. Unfortunately, resonance scattering tends to become more and more important with increasing energy.† Therefore, the method is useful, primarily, for the treatment of low-lying resonances. We are inclined to rely on the superposition method only in computations covering energies in the range $0.625 \text{ eV} \leq E \leq 100 \text{ eV}$, but actually this particular limitation of the superposition method is not as serious as it may seem. It is in the low-energy band of the resonance range that Monte Carlo is most helpful. Most often it is in this range that resonances overlap and shield each other, and, in this range, flux shapes may become extremely complicated. At higher energies, on the other hand, approximate deterministic methods, such as Nordheim's (Ref. 1), are usually perfectly adequate. The superposition method and Nordheim's method can be combined into a computational scheme which is both fast and accurate, one which can be used to design reactors or, if the need arises, to analyze experiments involving isolated fuel rods.

In closing we should like to restate the purpose of much of our work in these last three chapters. We have discussed a number of special techniques

---

† This is true because the neutron width is proportional to the square root of E, while the gamma width is almost energy independent.

for solving certain special problems. The problems are important in themselves, but just as important is a principle which we have tried to demonstrate. We have tried to show that the Monte Carlo method is not simply a heavy bludgeon, but a delicate, versatile tool to be handled with care.

REFERENCES

1. L. W. NORDHEIM and G. KUNCIR, "A Program of Research and Calculations of Resonance Absorption," GA-2527 (August 28, 1961).
2. LAWRENCE DRESNER, "Resonance Absorption in Nuclear Reactors," Pergamon Press, New York, Oxford (1960).

Errata

| Page 20 | Line 8 | For: $\alpha_1 - \dfrac{\alpha \sqrt{\alpha_2 - \alpha_1^2}}{\sqrt{n-1}}$ | Read: $\alpha_1 - \dfrac{a \sqrt{\alpha_2 - \alpha_1^2}}{\sqrt{n-1}}$ |

Page 21 Line 17 For: t_i^n Read: t_{i_n}

Page 25 Eq. (1.5.21) For: $\phi(\rho) = \sup\limits_{F(x_i)<\rho} x_i$ Read: $\phi(\rho) = \sup\limits_{F(x)<\rho} x$

Page 33 Eq. (1.7.16) For: $= \dfrac{\iint\limits_{B_2} dx_1 dx_2}{\pi/4}$ Read: $= \dfrac{\iint\limits_{B_i} dx_1 dx_2}{\pi/4}$

Page 33 Line 4 For: $B_1 = \{(x_1, x_2) \mid h(x_1, x_2) < z,\ 0 < x_1,\ x_2 \leq 1\}$

Read: $B_1 = \big\{(x_1, x_2) \mid h(x_1, x_2) < z,$

$$x_2 < \sqrt{1 - x_1^2},\ 0 \leq x_1,\ x_2 \leq 1\big\}$$

Page 33 Eq. (1.7.19) For: ρ random. Read: ρ uniform on $[0, 1]$.

Page 44 Line 7 For: $(P_1, \ldots, P_n)$ Read: $(P_1, \ldots, P_N)$

Page 45 Eq. (2.2.11) For: $\sum_t^i \phi^i = \sum\limits_{j=1}^{G} \Sigma_t^i C^{ij} \phi^i + Q^i$

Read: $\sum_t^i \phi^i = \sum\limits_{j=1}^{G} \Sigma_t^j C^{ij} \phi^j + Q^i$

227

Page 46	Line 12	For: $p_j = 1 - \dfrac{\Sigma_a^j}{\Sigma_t^j}$	Read: $P_j = \dfrac{\Sigma_a^j}{\Sigma_t^j}$
Page 46	Line 15	For: Q_i	Read: Q^i
Page 46	Line -5	For: Q_i	Read: Q^i
Page 49	Eq. (2.3.8)	For: $P_{i_1}^\alpha$	Read: $p_{i_1}^1$
Page 50	Theorem 2.2	For: The random variable η_j	
		Read: The random variable η_{j_0}	
Page 51	Eq. (2.3.13)	For: δ_{j0}	Read: δ_{j_0}

* Because i_1 now is the state immediately <u>after</u> j_0

Page 53	Eq. (2.3.21)	For: $\cdots \dfrac{\Sigma_t^{i_k}}{\sum_{i=1}^N \Sigma_s^{i_k i}}$	Read: $\cdots \dfrac{\Sigma_t^{i_k}}{\sum_{i=1}^N \Sigma_s^{i_{k-1} i}}$
Page 54	Line 11	For: Q_i	Read: Q^i
Page 55	Eq. (2.3.25)	For: $, a)$	Read: $, \mathbf{a})$
Page 57	Line 2	For: Eq. (2.2.4)	Read: Eq. (2.4.4)
Page 57	Line -10	For: particle sentering	Read: particles entering
Page 59	Eq. (2.4.17)	For: $\nabla \cdot \boldsymbol{\omega}$	Read: $\boldsymbol{\omega} \cdot \nabla$
Page 60	Eq. (2.4.23)	For: $\nabla \cdot \boldsymbol{\omega}$	Read: $\boldsymbol{\omega} \cdot \nabla$

Page 62   Line –6   For: $\sum_{\alpha} P(\alpha)\xi(\alpha) = \sum_{k=0}^{\infty} \cdots$

Read: $\sum_{\alpha} P(\alpha)\xi(\alpha) = \sum_{k=1}^{\infty} \cdots$

Page 63   Line 2   For: Theorem 3.5   Read: Theorem 3.3

Page 64   Eq. (2.5.16)   For: $d\mathbf{r}$ Read: $d\mathbf{r}'$

Page 65   Eq. (2.5.18)   For: $\prod_{j=1}^{m}$ Read: $\prod_{j=1}^{m-1}$

Page 65   Line 12   For: $\prod_{j=1}^{m}$ Read: $\prod_{j=1}^{m-1}$

Page 65   Line 18   For: $\prod_{j=1}^{m}$ Read: $\prod_{j=1}^{m-1}$

Page 68   Line 4   For: first m Read: first $m - 1$

Page 69   Eq. (2.6.12)   For: $\Sigma_t(x_m)$ Read: $\Sigma_r(x_m)$

Page 69   Eq. (2.6.13)   For: Thus...

Read: Thus, $W_m^* = \prod_{j=1}^{m-1} w_j^*$, where

$$w_j^* = \begin{cases} \dfrac{\Sigma_{s,\delta}}{\Sigma_{s,\delta} + \alpha\Sigma_a} & \text{if collision } j \\ & \text{is a delta event} \\ \dfrac{\Sigma_s}{\Sigma_s + (1-\alpha)\Sigma_a} & \text{if collision } j \\ & \text{is a true event} \end{cases}$$

Page 74   Line –14   For: $E\left[\eta_r^*|d\right] = C_1 \dfrac{\Sigma_r}{\Sigma_t^*} \cdots$

Read: $E\left[\eta_r^*|\alpha\right] = C_1 \dfrac{\Sigma_r}{\Sigma_t} \cdots$

Page 74   Line –10   For: estimators $\dfrac{\Sigma_r}{\Sigma_t^*}$ Read: estimators $\dfrac{\Sigma_r}{\Sigma_t}$

Page 75   Line 6   For: $W_m^* = \dfrac{\Sigma_{s,\delta} + \Sigma_s}{\Sigma_t^*}$

$\qquad\qquad\qquad\qquad$ Read: $W_j^* = \dfrac{\Sigma_{s,\delta} + \Sigma_s}{\Sigma_t^*}$ (cf. p. 69)

Page 75   Lines –4, –3   For: which would be continuously removed by absorption in

$\qquad\qquad\qquad\qquad$ Read: which survives absorption when

Page 76   Line 9   For: and (2.5.27) discussed

$\qquad\qquad\qquad\qquad$ Read: and (2.6.27) discussed

Page 86   Eq. (3.3.3)   For: $\displaystyle\int_{-\infty}^{P} \cdots \int_{-\infty}^{P}$ Read: $\displaystyle\int_{-\infty}^{P_n} \cdots \int_{-\infty}^{P_1}$

Page 86   Footnote §   For: See first footnote on p. 92.

$\qquad\qquad\qquad\qquad$ Read: See first footnote on p. 42.

Page 87   Line –9   For: $\prod_{i-1}^{k-1}$ Read: $\prod_{i=1}^{k-1}$

Page 90   Line 6   For: $\prod_{i=2}^{N}$ Read: $\prod_{i=1}^{N}$

Page 91   Line 3   For: $\mathcal{K}S + \mathcal{K}_1 S + \cdots$ Read: $S + \mathcal{K}_1 S + \cdots$

Page 98   Line 13   For: ξ_N has variance   Read: $\overline{\xi_N}$ has variance

Page 102   Line 4   For: $\displaystyle\int_{1}^{1}$ Read: $\displaystyle\int_{0}^{1}$

Page 104   Line 9   For: $dP_1 \cdots P_k$ Read: $dP_1 \cdots dP_k$

Page 104 Line 14 For: $\displaystyle\prod_{j=2}^{i} K(P_j, P_{j-i1})$ Read: $\displaystyle\prod_{j=2}^{i} K(P_j, P_{j-1})$

Page 104 Line −8 For: $dP_1 \cdots P_i$ Read: $dP_1 \cdots dP_i$

Page 104 Line −4 For: $\displaystyle\times \prod_{j=1}^{k} \cdots$

Read:

$$\times \prod_{j=1}^{k-1} \left[q(P_j)/c(P_j) \right] dP_{i+1} \cdots dP_k \times q(P_k)$$
$$= 1 - \lim_{k\to\infty} \int_\Gamma \cdots \int_\Gamma K(P_k, P_{k-1}) \ldots K(P_{i+1}, P_i)$$
$$\times \prod_{j=1}^{k-1} \left[q(P_j)/c(P_j) \right] dP_{i+1} \cdots dP_k \times q(P_k)$$

Page 105 Line 12 For: in as much as Read: inasmuch as

Page 105 Line −2 For: (Eqs. 2.4.29, 2.6.3) Read: (Eq. 2.6.3)

Page 107 Line 5 For: characteristic function

Read: characteristic function†

Page 107 Line 10 For: r to R Read: r to A

Page 107 Line 12 For: as a sum of Read: as a product of

Page 107 Line −1 Add Footnote: † $\chi_R(P) = \begin{cases} 1 \text{ if } P \in R \\ 0 \text{ if } P \notin R \end{cases}$

Page 109 Lines 8 For: $c(E', \mathbf{r})$ Read: $c(\mathbf{E}', \mathbf{r})$

Page 109 Line 15 For: $\omega' \to \omega$ Read: $\omega' \to \omega_0$

Page 109 Line 16 For: $c(E', \mathbf{r})$ Read: $c(\mathbf{E}', \mathbf{r})$

Page 109 Lines 17 For: $\omega' \to \omega$ Read: $\omega' \to \omega_0$

Page 109   Line 17   For: $\Sigma_t (\mathbf{r} + s\omega) P(E' \cdots$

                                Read: $\Sigma_t (\mathbf{r} + s\omega) ds] P(E' \cdots$

Page 109   Line 18   For: $c(E', \mathbf{r})$ Read: $c(\mathbf{E}', \mathbf{r})$

Page 113   Eq. (3.7.7)   For: $c(P)$ Read: $c(P')$

Page 114   Line −4   For: $dP_1 \cdots P_k$ Read: $dP_1 \cdots dP_k$

Page 116   Eq. (3.7.13)   For: $\int \psi^*(P) S(P)$ Read: $\int \psi^*(P) S(P) dP$

Page 118   Eq. (3.7.20)   For: $\int_\Gamma K(P_{l-1}, P_l)$ Read: $\int_\Gamma K(P_{l-1}, P_l) dP_l$

Page 131   Line 17   For: (3.7.6)   Read: (3.7.7)

Page 132   Eq. (3.9.7)   For: $E\{(\hat{\xi} - E[\xi]) \cdots$ Read: $E\{(\hat{\xi} - E[\hat{\xi}]) \cdots$

Page 133   Line −7   For: $C \le [\cdots$ Read: $C \le \frac{1}{2}[\cdots$

Page 135   Eq. (3.11.3)   For: $c = \dfrac{V_1 - V_{12}}{\cdots}$ Read: $c = \dfrac{V_2 - V_{12}}{\cdots}$

Page 143   Line 6   For: $P_{I \to II}$ Read: $P_{II \to I}$

Page 145   Line −12   For: $\int_0^l e^{-\Sigma_{aII} l} dt$ Read: $\int_0^l e^{-t \Sigma_{aII}} dt$

Page 147   Eq. (4.3.7)   For: $RA \equiv \int_I = \cdots = \int_{II} \cdots$

                                Read: $RA \equiv \int_{II} = \cdots = \int_I \cdots$

Page 154   Line 10   For: $= \dfrac{1}{3}$ Read: $= \dfrac{1}{3} \dfrac{1}{(2\Sigma_a t)^2}$

Page 155 Table 4.1, $\rho = 1.4$ For: 1.1389 Read: 3.1389

Page 155 Table 4.1, $\rho = 2.4$ For: 4.8450 Read: 4.9841

Page 155 Table 4.1, $\rho = 3.8$ For: 3.7059 Read: 7.7059

Page 157 Footnote † For: l_{II} Read: $\overline{\overline{l}}_{\mathrm{II}}$

Page 159 Eq. (4.6.7) For: $\overline{W} = \cdots = \cdots$

Read: $\overline{W} = \cdots A_{\mathrm{II}} = \cdots A_{\mathrm{II}}$

Page 160 Eq. (4.7.3) For: $F_{TU}(\mathbf{r}, \omega) + \dfrac{1}{4\pi}Q_{\mathrm{I}}$

Read: $F_{TU}(\mathbf{r}, \omega') + \dfrac{1}{4\pi}Q_{\mathrm{I}}$

Page 160 Eq. (4.7.5) For: $\ldots \Sigma_{T\mathrm{II}}.$

Read: $\ldots \Sigma_{T\mathrm{II}}, \ \mathbf{r} \in R_{\mathrm{II}}.$

Page 161 Eq. (4.7.9) For: $-F_\beta(\mathbf{r}, \omega).$

Read: $-F_\beta(\mathbf{r}, \omega), \ \mathbf{r} \in R_{\mathrm{II}}.$

Page 166 Line 6 For: $\phi_{\beta u} \cdots$ Read: $\phi_{\beta u r} \ldots$

Page 170 Eq. (4.10.1) For: $d\omega$ Read: $d\omega'$

Page 170 Line 11 For: $\phi(\mathbf{r} \cdot \omega)$ Read: $\phi(\mathbf{r}, \omega)$

Page 172 Line 18 For: $\phi^*(r) =$ Read: $\phi^*(\mathbf{r}) =$

Page 179   Line −11   For: $W_1 = \dfrac{E_1 - E_2}{2}$

Read: $W_1 = E_1 - E_2$

Page 180   Line 18   For: $d'\omega'$ Read: $d\omega'$

Page 180   Eq. (5.2.9)   For: $\omega : \mathbf{r})d\omega'$ Read: $\omega, \mathbf{r})d\omega'$

Page 181   Line −7   For: $\mu_0 = 0$ Read: $\mu_0 = 1$

Page 182   Lines 1   For: $\delta(\mu_0)$ Read: $\delta(\mu_0 - 1)$

Page 182   Line 3   For: $\delta(\mu_0)$ Read: $\delta(\mu_0 - 1)$

Page 182   Line 3   For: $\delta_1(\mu_0)$ Read: $\delta_1(\mu_0 - 1)$

Page 182   Line 4   For: $\delta_1(\mu_0)$ Read: $\delta_1(\mu_0 - 1)$

Page 183   Line −8   For: $k_{ij}(\omega' \cdot \omega)$ Read: $c_{ij}(\omega' \cdot \omega)$

Page 186   Line 3   For: $\displaystyle\int_V$ Read: $\displaystyle\int_R$

Page 190   Line 16   For: $I = \displaystyle\sum_{j=1}^{N}$ Read: $I = \displaystyle\sum_{i=1}^{N}$

Page 192   Lines 2–3   For: $\dfrac{P_{\mathrm{I}\to\mathrm{II}}}{1 - P_{\mathrm{I}\to\mathrm{II}}}$ Read: $\dfrac{1 - P_{\mathrm{I}\to\mathrm{II}}}{P_{\mathrm{I}\to\mathrm{II}}}$

Page 192   Line 7   For: $\dfrac{P^*_{\mathrm{II}\to\mathrm{I}}}{1 - P^*_{\mathrm{II}\to\mathrm{I}}}$ Read: $\dfrac{1 - P^*_{\mathrm{II}\to\mathrm{I}}}{P^*_{\mathrm{II}\to\mathrm{I}}}$

Page 194   Line 4   For: $W_i\widetilde{k_{ij}}(\mu_0) =$ Read: $W_i\widetilde{k_{ij}}(\mu_0)$

Page 199   Line −5   For: in II   Read: in III

Page 199   Eq. (5.5.1)   For: $\overline{\phi}_I A_I = Q_{II} \int_{II} \phi^*(\mathbf{r}) d\mathbf{r}$

Read: $\overline{\phi}_I V_I = Q_{III} \int_{III} \phi^*(\mathbf{r}) d\mathbf{r}$

Page 200   Line 18   For: region II   Read: region III

Page 200   Line 19   For: region II   Read: region III

Page 208   Lines 4, 6, 8   For: $\int_{E_L}^{E_u}$ Read: $V_f \int_{E_L}^{E_u}$

Page 210   Lines −13, −12   For: l_i, $i-$th, $\sum_{i=1}^{m} l_i$

Read: l_j, $j-$th, $\sum_{j=1}^{m} l_j$

Page 218   Eq. (6.5.5)   For: Σ_{Sf} Read: Σ_S

Page 219   Line 5   For: $\phi_{\beta uu}(\mathbf{r})$ Read: $\phi_{\beta uu}(E, \mathbf{r})$

Page 219   Line 9   For: $I_0 = V_I \int_{E_L}^{E_u} dE' \Sigma_{af}(E) \cdots$

Read: $\Sigma_{af}(E')$

Page 220   Eq. (6.5.13)   For: Σ_{Sf} Read: Σ_S

Page 221   Line −11   For: draw   Read: drawn

Page 221   Line −1   For: $W_4 = \int_{E_L}^{E_u} Q_4(E') \text{sgn}[\Sigma_{Srf}(E')] dE'$

Read: $W_4 = \text{sgn}[\Sigma_{Srf}(E)] \int_{E_L}^{E_u} Q_4(E') dE'$

Page 223   Line −12   For: if $\Sigma_{Srf}(E) > 0$ Read: if $\Sigma_{rf}(E) > 0$

Author Index

Italicized page numbers refer to names cited in a reference list appearing at the end of each chapter.

237

Subject Index